Metallic Coatings for Corrosion Control

Corrosion Control series

This is the first of a series of monographs on practical aspects of corrosion control, under the general editorship of **L. L. Shreir**, PhD, FRIC, FIM, FICorrT, FIMF, editor of the standard reference work *Corrosion.*

Metallic Coatings for Corrosion Control

V. E. Carter, FICorrT, FIMF

NEWNES–BUTTERWORTHS
LONDON BOSTON
Sydney Wellington Durban Toronto

THE BUTTERWORTH GROUP

United Kingdom — Butterworth & Co (Publishers) Ltd
London: 88 Kingsway, WC2B 6AB

Australia — Butterworths Pty Ltd
Sydney: 586 Pacific Highway, Chatswood, NSW 2067
Also at Melbourne, Brisbane, Adelaide and Perth

Canada — Butterworth & Co (Canada) Ltd
Toronto: 2265 Midland Avenue, Scarborough, Ontario, M1P 4S1

New Zealand — Butterworths of New Zealand Ltd
Wellington: 26-28 Waring Taylor Street, 1

South Africa — Butterworth & Co (South Africa) (Pty) Ltd
Durban: 152-154 Gale Street

USA — Butterworth (Publishers) Inc
Boston: 19 Cummings Park, Woburn, Mass. 01801

First published in 1977

ISBN 0 408 00270 0

Typeset by G. A. Pindar and Son Ltd., PO Box 8, Newlands Park, Scarborough, Yorks. Printed in England by Chapel River Press, Andover, Hants.

Preface

On setting out to write a book on metallic coatings for corrosion control the problem that faced me was not what to include but rather what to omit in order to provide a coverage of the subject that would be sufficiently comprehensive within the limitations imposed by one reasonably sized volume. Similarly, I had to decide what depth of treatment should be given to each aspect of the subject matter included. The series of monographs on corrosion science and technology, in which this book is included, is intended to provide volumes covering practical aspects of corrosion control that will be of use to engineers who have to combat corrosion when they design, make and operate structures and other equipment.

I decided, therefore, to attempt to show in broad outline the ways in which metal coatings can control corrosion and the various ways in which such coatings can be applied. In order to select from the wide range of coatings the one best suited to a particular problem of corrosion control, it is necessary to understand what is involved in the preparation and application of each of the available methods. It is also necessary to understand how the limitations of individual steps in a given process can affect the quality and performance of the finished article. On the other hand, I feel that the practical reader of this book will neither need nor wish to be involved too deeply in the more technical aspects of the coating processes; nor should such a book be expected to enable the reader to 'do the whole job for himself'. In this, as in so many fields, the need is to have sufficient understanding of the problem to be able to select the answer appropriate to the need and then employ the experts in that technique. Finally, when a metal coating process has been chosen and applied, it is necessary to test its quality and performance.

For those readers who wish to delve further into the detail of any particular aspect of metal coatings there is a wide range of specialised books that can be consulted. I have limited myself to the barest minimum of literature references in the text, and have added to the list of these references at the end of the book a short bibliography of works that can be consulted by those whose interests lie in a particular direction. There are, of course, a host of other books available, any of which might equally well have been included in this list, and my selection is not intended to imply superiority over any that have been omitted.

I am greatly indebted to Dr L.L. Shreir for helpful discussions concerning the form and content of this book. I would also like to thank him for his kind assistance in the scientific editing of the book and for his help in the preparation of Chapter 1.

I thank the Director of the BNF Metals Technology Centre for very kindly supplying the photographs that appear in the book and for granting permission for them to be reproduced.

V.E.C.

Contents

1

Metallic corrosion

Thin metallic coatings are applied to substrates (metals, plastics, etc.) for a variety of reasons (see Chapter 2), but the major application is undoubtedly corrosion protection. This provides an economical means of combining the properties of the substrate and the metallic coating to give a composite material that has both good mechanical properties and good corrosion resistance. Thus mild steel has excellent mechanical properties, is easily fabricated and is cheap, but its resistance to corrosion in most environments is poor, and the rusting of steel results in progressive deterioration of the structure or component. This disadvantage can be overcome by alloying the steel with the more corrosion-resistant metals nickel and chromium to give the rust-resistant 18Cr–8Ni austenitic stainless steel, but alloys of this type are relatively expensive. A more economical approach is to apply a thin coating of nickel followed by an even thinner coating of chromium, a procedure widely used for producing corrosion-resistant decorative finishes that have the mechanical properties of mild steel and the corrosion resistance of chromium and nickel. Zinc has excellent resistance to corrosion in a variety of environments, including the atmosphere and natural waters, but it is mechanically weak, difficult to fabricate and relatively expensive; it is, of course, used for roofing sheet and flashings, applications in which mechanical strength is relatively unimportant. However, it can be readily applied to mild steel by hot dipping, electroplating, metal spraying or high-temperature diffusion, and coatings applied by these methods are widely used for protecting a variety of steel structures and components where appearance is a secondary consideration.

Ideally, a metallic coating applied to an alloy such as mild steel should form a continuous barrier that completely isolates the underlying metal from the environment. Unfortunately, this is seldom possible in practice since the method of application of the coating gives rise to discontinuities such as pores, pits and cracks. Furthermore, discontinuities may be produced during the subsequent forming operations — for example, cut edges — or by mechanical damage or removal of the coating by corrosion during actual service. Thus coatings are usually discontinuous, and it is therefore necessary to consider not only the corrosion resistance of the substrate and the coating but their effects on one another when they are in contact. This *bimetallic corrosion,* or the effect on the corrosion rates of two dissimilar metals when in contact, has certain unusual features in the case of metallic coatings on a metallic substrate, owing to the relatively small areas of the substrate metal that are exposed to the environment through discontinuities in the coating.

Nature of corrosion

Metals, with the exception of the noble metals Cu, Ag, Au, Hg and the Pt metals, are usually found in nature combined with non-metals as oxides, silicates, carbonates, sulphides, etc., and since these have existed in the earth from time immemorial it follows that these compounds, rather than the metal, are the stable form of the metal. To obtain the metal the ore (in which the compound of the metal, or mineral, is present in a suitable form and at a sufficient concentration for its conversion to the metal to be technologically and economically feasible) is subjected to a reduction process, in which energy is supplied to the system in the form of chemical, electrical or thermal energy. For example, the oxides of zinc and iron can be reduced to the metal by using the chemical energy of carbon, which has a greater affinity for oxygen than the metal, and the process can be represented by the chemical reactions:

$$2ZnO + C \rightarrow 2Zn + CO_2 \tag{1.1}$$

$$2Fe_2O_3 + 3C \rightarrow 4Fe + 3CO_2 \tag{1.2}$$

Alternatively, in the case of zinc, the oxide can be leached from the ore by means of sulphuric acid, and the acid solution of zinc sulphate can then be electrolysed (supply of electrical energy) to give a deposit of zinc metal at the cathode:

$$ZnSO_4 + H_2O \rightarrow Zn + H_2SO_4 + \tfrac{1}{2}O_2 \tag{1.3}$$

The overall reaction (equation 1.3) can also be written in the form of two half-reactions showing the cathodic reduction of zinc ions to zinc metal and the anodic oxidation of water to oxygen and acid (hydrogen ions):

$$\text{Cathodic reduction} \quad Zn^{2+} + 2e \rightarrow Zn \qquad (1.4)$$

$$\text{Anodic oxidation} \quad H_2O \rightarrow 2H^+ + \tfrac{1}{2}O_2 + 2e \qquad (1.5)$$

It should be noted that a similar process is used for electroplating zinc, but whereas in the extraction process an inert anode is used (frequently lead) a zinc anode is used in electroplating in order to ensure that the concentration of zinc ions is maintained fairly constant; the anodic reaction in this case is the reverse of equation 1.4, the zinc metal being oxidised to zinc ions.

The fact that energy is consumed in the reduction process means that the metal has a higher energy state than the metal compound; this in turn means that the metal is unstable and tends to revert to the

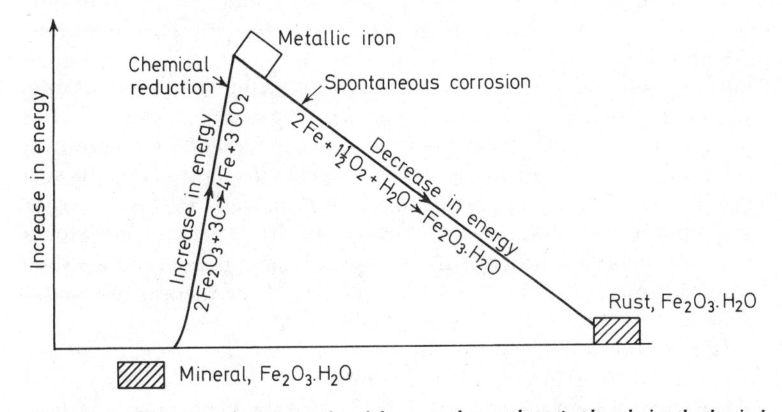

Figure 1.1 Diagrammatic representation of the energy changes that take place during the chemical reduction of a mineral (hematite, $Fe_2O_3.H_2O$) to the metal (iron), and the subsequent spontaneous conversion of the metal back to the oxide (corrosion product) during exposure to water and oxygen

combined form when it comes into contact with non-metals in the environment. This sequence of events in which the mineral is reduced to the metal, with a consequent increase in energy, and the subsequent spontaneous conversion of the metal back to a compound (referred to as a *corrosion product,* which frequently has the same composition and crystal structure as the mineral) with a decrease in energy is illustrated in *Figure 1.1*. Thus when iron is exposed to an aqueous environment containing dissolved oxygen, which will occur when it is exposed to the atmosphere or immersed in a natural water, it tends to revert to its oxide $Fe_2O_3.H_2O$, *rust*, which corresponds in composition to the naturally occurring mineral *lepidocrocite.* Indeed if metal extraction is regarded as the winning of a metal from its ores for a profit then corrosion is the converse process, i.e. loss of metal and loss of profit.

Definition of corrosion

Metals are used for engineering constructions because of their mechanical properties, and clearly these are affected by the conversion of the metal into corrosion products, although as will be seen this obviously depends on the *rate* of corrosion and the *extent* to which it has proceeded. Metals are also used because of their aesthetic appeal, and although the formation of thin surface films (tarnish films) has little effect on mechanical properties they are detrimental to appearance and to other properties; a black sulphide tarnish on silver is aesthetically undesirable and also has a detrimental effect on properties if the silver is used, for example, as an electrical contact.

A precise definition of corrosion is not as simple as might be thought, since implicit in this term are a number of different concepts, but for the purpose of this work the following definition will be adopted: *reaction of a metal or alloy with its environment with the formation of corrosion products.* In so far as any conversion of a metal to its corrosion products must be regarded as detrimental to the metal it could be said that corrosion is always detrimental, but this depends on the *rate* of corrosion and its *extent* — in many systems the rate may be so slow that its effect is negligible, in others it may be significant but tolerable, while in others it may be so high that its consequences are catastrophic.

It is important to emphasise that whereas the mechanical and physical properties of metals and alloys are independent of the nature of the environment the converse applies to their corrosion properties. Thus in specifying the tensile strength of a mild steel of given composition it is not necessary to specify the environmental conditions prevailing during the actual determination, which in any case requires only a few minutes in a tensile-testing machine. However, the corrosion rate of mild steel obviously depends on the environment and, for example, is far more rapid in an industrial polluted atmosphere than in a clean rural atmosphere — it is also more rapid in sea water than in a fresh potable water.

It follows that the corrosion of a given metal, i.e. the corrosion rate, depends upon the precise environmental conditions to which it will be subjected in service. As far as different metals and alloys are concerned each has a specific corrosion rate in a given environment; for example, whereas mild steel corrodes rapidly when exposed to an industrial atmosphere, a stainless steel of the type 18Cr–10Ni–3Mo will be virtually uncorroded and its surface remain bright and reflecting. *Figure 1.2* illustrates the complexity of corrosion and shows

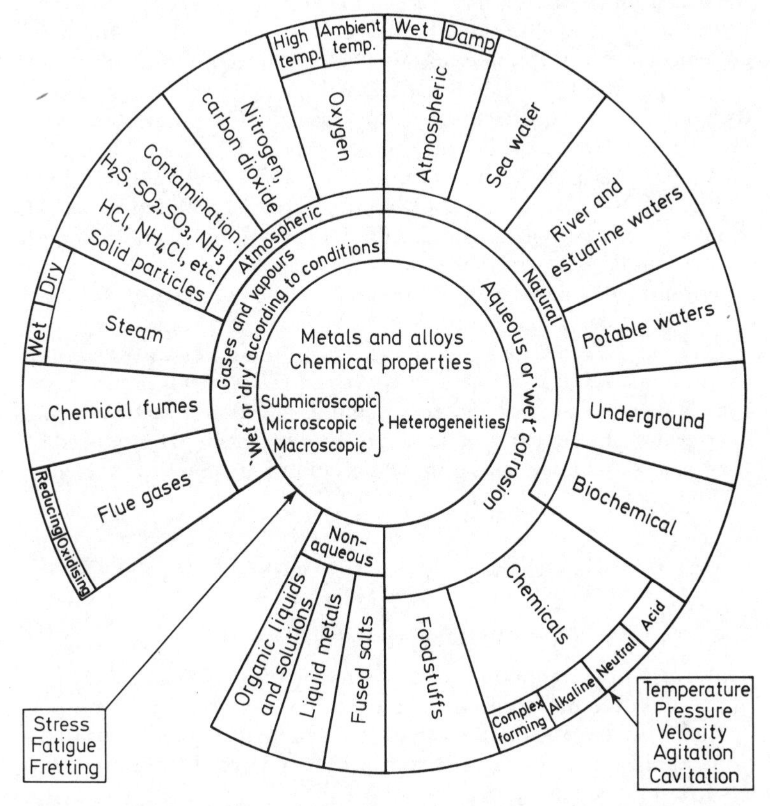

Figure 1.2 Illustration of how the corrosion of a metal (rate and form of attack) is dependent on the nature of the environment and the metal and on environmental conditions and stress (From Corrosion, ed. L.L. Shreir, Newnes–Butterworths, 1976)

that it depends not only upon the environment (chemical composition) and environmental conditions (temperature, pressure, velocity, agitation, etc.) but also upon the composition and structure of the alloy, and upon stress (type and magnitude).

Metal coatings are used to protect the substrate from corrosion in a variety of environments that range from acid fruit juices to high-temperature combustion products, and each environment presents special problems. However, the major application for coatings is for the protection of structures and components that are in contact with natural aqueous environments, which include various atmospheres (industrial, rural, marine; outdoor and indoor exposure) and waters (fresh, brackish, sea water, polluted water).

As far as the atmosphere is concerned it should be noted that, with the exception of the sulphide tarnishing of silver and copper, the presence of water on the metal surface is an essential condition for aqueous corrosion and that this may range from a thin condensed film of moisture resulting from fluctuations in temperature to completely wet conditions resulting from heavy rainfall. Contaminants in the atmosphere can markedly affect the corrosion rate, and this applies particularly to gases such as SO_2 and H_2S and to solid particles such as carbon, NH_4Cl and $(NH_4)_2 SO_4$; in general, a polluted industrial atmosphere is far more corrosive than a rural atmosphere. It should also be noted that the atmospheres in locations near the sea are contaminated with particles of salt, and this too has a significant effect on the corrosion rate.

Different waters vary significantly in their corrosiveness; for example sea water is more corrosive than fresh natural waters. However, hard waters (fresh or saline) contain calcium bicarbonate and magnesium sulphate in solution, and the increase in pH produced by the cathodic reaction results in the precipitation of insoluble calcium carbonate and magnesium hydroxide:

$$CaCO_3 + H_2CO_3 \underset{\text{increase in pH}}{\overset{\text{decrease in pH}}{\rightleftharpoons}} Ca(HCO_3)_2$$

$$MgSO_4 + 2NaOH \rightarrow Mg(OH)_2 + Na_2SO_4$$

These insoluble compounds, if deposited on the surface of a corroding metal or within discontinuities in a coating, can act as a barrier that partially isolates the metal from the environment. Similar considerations apply to insoluble corrosion products formed from the metal substrate or metal coating.

Forms of corrosion

If corrosion is uniformly distributed over the metal surface and if the rate is assumed to follow a linear law, it is possible to define corrosion in terms of a weight loss per unit area per unit time. Various units are used for this purpose but the two most important are:

Milligrams per square decimeter per day	$mg\ dm^{-2}\ d^{-1}$
Grams per square metre per day	$g\ m^{-2}\ d^{-1}$

Furthermore, these units may be converted into rates of penetration if the density of the metal is taken into account; typical units are millimetres per year and inches per year.

Corrosion if uniform may be defined in terms of a rate, and this provides a means of predicting the depth of penetration into the metal after any predetermined time; although uniform corrosion is obviously detrimental it is at least predictable, and allowances for corrosion are frequently made in the design of structures. Uniform (or

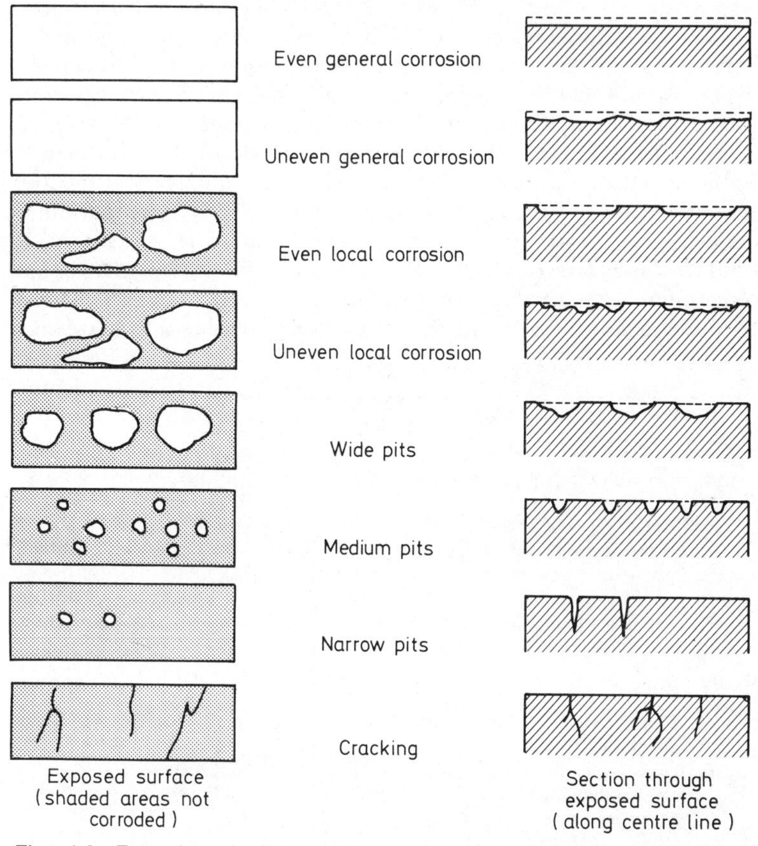

Figure 1.3 Forms of corrosion (From Corrosion, ed. L.L. Shreir, Newnes–Butterworths, 1976)

near-uniform) corrosion occurs during the corrosion of a metal in acid and during the exposure of certain metals to natural environments. However, in many metal/environment systems attack is localised and, although the whole of the metal surface corrodes, certain areas corrode somewhat more rapidly than others, giving rise to shallow saucer-shaped areas of attack. At the other extreme the major part of the surface of the metal remains almost unattacked while certain small areas are attacked at a very high rate, with rapid penetration into the section of the metal, which can lead to complete perforation when the cross-section is comparatively thin. (See *Figure 1.3.*) This type of attack in which the diameter of the area of surface attacked is comparable to or smaller than the depth of penetration is referred to as *pitting,* and is generally regarded as more insidious than uniform attack since the sites of attack and the rate of penetration into the metal are often unpredictable. It is evident that a weight-loss determination would give misleading information, since although the weight loss would be small the perforation of a metal used as a container for a fluid could result in costly and serious damage.

In general, attack is uniform when the metal/environment system is homogeneous, i.e. the metal is uniform in composition and the nature (composition, oxygen concentration, pH, etc.), temperature, velocity, etc., of the environment is the same at all parts of the metal surface. Conversely, heterogeneities in the metal and/or the environment tend to give rise to localised attack, although the intensity of attack will depend on the system under consideration. As far as

Table 1.1 HETEROGENEITIES IN METALS

1. *Atomic* (as classified by Ehrlich and Turnbull, *Physical metallurgy of stress corrosion fracture,* Interscience, 47 (1959))
   - (*a*) Sites within a given surface layer ('normal' sites); these vary according to particular crystal plane
   - (*b*) Sites at edges of partially complete layers
   - (*c*) Point defects in the surface layer: vacancies (molecules missing in surface layer), kink sites (molecules missing at edge of layer), molecules adsorbed on top of complete layer
   - (*d*) Disordered molecules at point of emergence of dislocations (screw or edge) in metal surface

2. *Microscopic*
   - (*a*) Grain boundaries—usually, but not invariably, more reactive than grain interior
   - (*b*) Phases—metallic (single metals, solid solutions, intermetallic compounds), non-metallic, metal compounds, impurities, etc.—heterogeneities due to thermal or mechanical causes

3. *Macroscopic*
   - (*a*) Grain boundaries
   - (*b*) Discontinuities on metal surface—cut edges, scratches, discontinuities in oxide films (or other chemical films) or in applied metallic or non-metallic coatings
   - (*c*) Bimetallic couples of dissimilar metals
   - (*d*) Geometrical factors—general design, crevices, contact with non-metallic materials, etc.

the metal or alloy is concerned, grain boundaries, different phases and different mechanical or thermal treatments are heterogeneities that favour localised attack; the metallographic etching of a metal to reveal structure is based on this principle, and since grain boundaries are attacked at a higher rate than grain interiors they appear as a dark network when examined microscopically. Similar considerations apply to those grains whose orientation is such that crystal faces that corrode at the highest rate are exposed at the surface. *Table 1.1* shows how heterogeneities in the metal or environment may lead to attack being concentrated on one area (the anode) while the remainder of the surface is unattacked (the cathode).

Electrochemical mechanism of corrosion

The spontaneous corrosion of metals in aqueous solutions and the electrodeposition of metals from aqueous solutions of their salts are characterised by the fact that they are both electrochemical. For this reason it is appropriate to consider them in this chapter, although emphasis will be placed on bimetallic corrosion, which is of particular importance in the behaviour of a discontinuous metallic coating on a less corrosion-resistant metal substrate.

Electrochemical cells

An electrochemical cell is a device for converting the chemical energy of a spontaneous chemical reaction into electrical energy (work energy) and heat. This can be exemplified by the Daniell cell, which utilises the chemical energy of the reaction

$$CuSO_4 + Zn \rightarrow Cu + ZnSO_4 \qquad (1.6)$$

which if allowed to proceed in a beaker would be converted solely into heat energy. *Figure 1.4* shows the cell, which can be represented by

$$Zn|Zn^{2+}(aq.)|Cu^{2+}(aq.)|Cu$$

and which if connected across the terminals by a metallic conductor results in the transfer of electrical charge (electrons) at the two interfaces (electrode reactions), transfer of charge (electrons) through the metallic circuit (electronic conduction), and transfer of charge (cations and anions) through the solution (electrolytic conduction).

At the Zn/Zn^{2+} electrode positive charge (Zn^{2+}) is transferred from the metal electrode to the solution (or, alternatively, this could be regarded as the transfer of negative charge or electrons in the reverse direction), and the half-reaction is

$$Zn(l) \rightarrow Zn^{2+}(aq.) + 2e \qquad (1.7)$$

where Zn(l) represents a zinc ion in the lattice of the metal and Zn^{2+}(aq.) represents a hydrated zinc ion in solution. This loss of electrons represents the oxidation of zinc metal to a higher valence state,

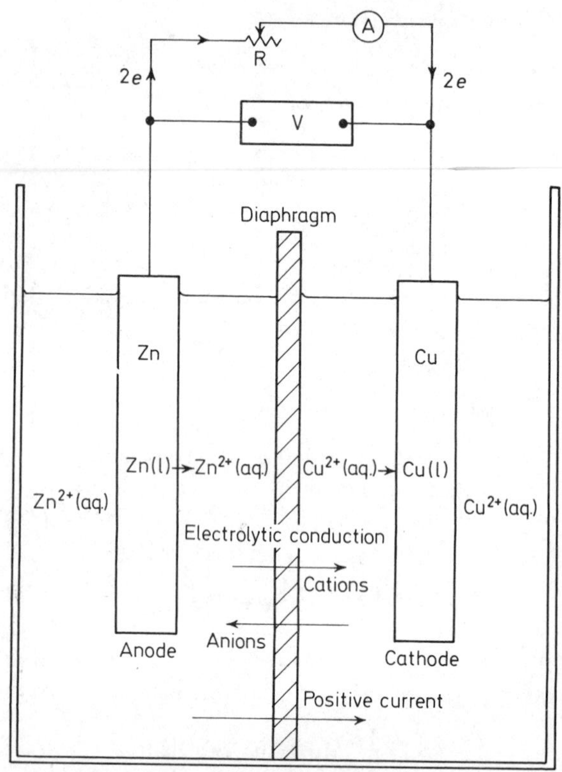

Figure 1.4 Daniell cell, in which the chemical energy of the spontaneous reaction $Cu^{2+} + Zn \rightarrow Cu + Zn^{2+}$ is converted into electrical energy. In this electrochemical reaction the zinc is anodically oxidised to Zn^{2+}(aq.) and the Cu^{2+}(aq.) ions in solution are cathodically reduced to copper. Note that the rate and extent of the two half-reactions are interdependent and electrochemically equivalent

and by definition an electrode at which oxidation takes place is the *anode*. It is apparent that the zinc electrode corrodes and that its conversion to zinc ions releases energy in the form of electrical energy, although a proportion of this energy also appears as heat energy.

At the Cu/Cu^{2+} electrode positive charges (Cu^{2+}ions) are transferred from the solution to the electrode, and the acceptance of electrons by these ions results in their reduction to metal atoms:

$$Cu^{2+}\ (aq.) + 2e \rightarrow Cu(l) \qquad (1.8)$$

An electrode at which a species in solution accepts electrons and is reduced to a lower valence state is defined as the *cathode.*

There are two features of this cell that require detailed consideration: (a) the rates of the electrode reactions, and (b) the potentials of the electrodes and the e.m.f. of the cell.

Rates of reaction

The rate of an electrochemical reaction is given by the current I (unit amperes, A), and a current of 1 A is equivalent to a rate of 1 coulomb per second ($1\ C\ s^{-1}$). However, since a surface is involved the rate per unit area of surface is more significant than the rate, i.e. the current density $i = I/S$, where S is the area in appropriate units (cm^2, mm^2, m^2). The relationship between quantity of electrical charge and the extent of an electrode reaction is given by Faraday's Law, which states that 1 faraday of charge ≡ 1 gram equivalent of electrochemical charge, i.e.

$$1\ \text{faraday} = M/z \qquad (1.9)$$

where M is the molar mass (kg), z is the number of electrons involved in one act of the electrode reaction, and 1 faraday ≈ 96 500 C. Thus the rate of an electrochemical reaction k_e is given by

$$k_e = \frac{i}{zF}\ \text{mol cm}^{-2}\ \text{s}^{-1} \qquad (1.10)$$

where i is the current density in $A\ cm^{-2}$. If M is the molar mass in kg,

$$k_e = \frac{Mi}{zF}\ \text{kg cm}^{-2}\ \text{s}^{-1} \qquad (1.11)$$

It is evident that in any electrochemical reaction the rates of the anodic and cathodic reactions as given by equations 1.7 and 1.8 are interdependent, i.e. the rate and extent of the anodic reaction $Zn \rightarrow Zn^{2+}(aq.) + 2e$ must equal that of the cathodic reaction $Cu^{2+}(aq.) + 2e \rightarrow Cu$.

Potentials and e.m.f.

A metal consists of an orderly arrangement of metal cations surrounded by a cloud of free electrons so that any point within the metal is electrically neutral; a solution consists of hydrated cations and hydrated anions and again electroneutrality prevails. However, at the interface between two phases there is a redistribution of charge and this gives rise to an *electrical double layer,* which may be regarded as the two plates of a capacitor (*Figure 1.5*).

Thus when copper is immersed in a solution of its cations there is a tendency for aquo copper ions in solution to discharge and form

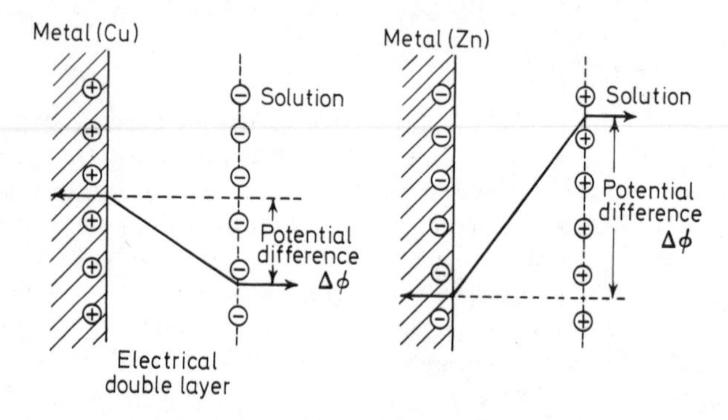

Figure 1.5 *Electrical double layer at the interfaces Cu/Cu^{2+} (aq.) and Zn/Zn^{2+} (aq.) resulting in a potential difference $\Delta\varphi$. When determined with a standard hydrogen electrode, and when $a_{Zn^{2+}} = a_{Cu^{2+}} = 1$, $\Delta\varphi = 0.34$ V for the copper electrode and -0.76 V for the zinc electrode. Note that within the metal and within the bulk solution the distribution of charge is uniform and that there is no potential difference*

copper ions in the lattice, and also a tendency for copper ions in the lattice to pass into solution and to form aquo copper ions in the solution. Initially the rates of these two processes are unequal, and in the case of the system Cu/Cu^{2+} the reaction Cu^{2+} (aq.) + $2e \rightarrow Cu(l)$ is the more rapid so that there is a net positive charge on the metal side of the interface, although the electrical double layer as a whole is electrically neutral. This results in a field in a direction perpendicular to the interface; this in turn results in the rates of the two reactions becoming equal. The system is now in a state of dynamic equilibrium in which the rate of Cu^{2+}(aq.) + $2e \rightarrow Cu(l)$ is equal to the rate of $Cu(l) \rightarrow Cu^{2+}$(aq.) + $2e$, and the potential is now the equilibrium potential $E_{eq.}$.

There are a number of requirements that are essential in the determination of the equilibrium potential difference at a single interface, e.g. the equilibrium potential difference at the interface

Cu/Cu^{2+}(aq.). It is not possible to measure the potential of a single interface; it must be coupled to another interface, thus producing an electrochemical cell whose equilibrium e.m.f. must be determined in such a way that the rate of the cell reaction is practically zero. It follows that a voltmeter of low resistance will not meet this requirement, and that it is essential to use a potentiometer or a high-impedance electrometer. It should be noted that the two metal/solution interfaces forming the cell are called half-cells and the reaction that occurs in each cell is called a half-reaction. In order to evaluate the potential difference of one half-cell from the measured e.m.f. it is necessary to give an arbitrary potential to a specific half-cell and half-reaction, and for this purpose the equilibrium between hydrogen ions and hydrogen gas

$$2H^+ + 2e \leftrightharpoons H^2 \quad (1.12)$$

has been selected. This equilibrium involves the transfer of electrons between the hydrogen ions and hydrogen gas, and this can be achieved only if an electronic conductor is immersed in the solution; for this purpose platinised platinum (platinum with a coating of finely divided platinum) is used, since it is a good catalyst for the forward and reverse reactions that constitute the equilibrium.

The Nernst equation

The potential difference, *E*, at an interface obviously depends on the nature of the system, but also on the concentration (or more precisely the activity, *a*) of the ions in solution and the pressure (or more precisely the fugacity) of the gas involved in the equilibrium. This relationship is given by the Nernst equation:

$$E = E^{\ominus} - \frac{RT}{zF} \ln \frac{a_{\text{products}}}{a_{\text{reactants}}} \quad (1.13)$$

where $E^{\ominus}$ is a constant (the *standard electrode potential*), *R* is the gas constant 8.314 J mol^{-1} and *T* is the temperature (K). Substituting for the constants and at a temperature of 25°C (298 K), and bearing in mind that $\ln x = 2.303 \log x$, equation 1.13 can be written as

$$E = E^{\ominus} - \frac{0.059}{z} \log \frac{a_{\text{products}}}{a_{\text{reactants}}} \quad (1.14)$$

It is apparent that the sign of $E^{\ominus}$ depends on the way in which the equilibrium is written. Thus for the equilibrium between Cu(l) and Cu^{2+}(aq.) the equilibrium can be written as

$$\underset{\text{reactant}}{Cu^{2+}(aq.) + 2e} \rightarrow \underset{\text{product}}{Cu(l)} \qquad (1.15)$$

or

$$\underset{\text{reactant}}{Cu(l)} \rightarrow \underset{\text{product}}{Cu^{2+}(aq.) + 2e} \qquad (1.16)$$

and both are equally correct, but it is evident that for the sign of E to be the same for a given activity of Cu^{2+} the sign of $E^{\ominus}$ must be positive for equation 1.15 and negative for equation 1.16. It is now an accepted convention that half-reactions must be written with the electrons on the left-hand side as in equation 1.15, i.e. as a reduction: $E^{\ominus}_{Cu^{2+}/Cu} = +0.34$ V and $E^{\ominus}_{Zn^{2+}/Zn} = -0.76$ V. If the activities of the reactants and products are unity the second term on the righthand side of equation 1.14 becomes zero, and $E = E^{\ominus}$; thus the standard electrode potential $E^{\ominus}$ is defined as the potential of the interface at unit activity of all species involved in the equilibrium. In the case of the hydrogen electrode the equilibrium is given by equation 1.12, and when $a_{H^+} = p_{H_2} = 1$, $E = E^{\ominus}$ and $E^{\ominus}$ is given the arbitrary potential of 0.00 V. This forms the arbitrary reference for all other potentials, which are thus expressed on the *hydrogen scale,* i.e. with reference to the Standard Hydrogen Electrode (SHE).

Table 1.2 gives the standard electrode potentials of the more common $M^{z+} + ze = M$ equilibria arranged in order of their potentials, and this table is also referred to as the *e.m.f. series of metals.* It is

Table 1.2 E.M.F. SERIES OF METALS (OR STANDARD ELECTRODE POTENTIALS OF $M^{z+} + ze = M$ EQUILIBRIA *vs* STANDARD HYDROGEN ELECTRODE

Equilibrium	*Standard electrode potential* (V) *vs SHE*
$Au^{3+} + 3e = Au$	1.50
$Ag^{+} + e = Ag$	0.799
$Hg_2^{2+} + 2e = 2Hg$	0.789
$Cu^{+} + e = Cu$	0.52
$Cu^{2+} + 2e = Cu$	0.337
$H^{+} + e = \frac{1}{2}H_2$	0.00
$Pb^{2+} + 2e = Pb$	−0.126
$Sn^{2+} + 2e = Sn$	−0.136
$Ni^{2+} + 2e = Ni$	−0.250
$Cd^{2+} + 2e = Cd$	−0.403
$Fe^{2+} + 2e = Fe$	−0.440
$Zn^{2+} + 2e = Zn$	−0.763
$Al^{3+} + 3e = Al$	−1.66
$Mg^{2+} + 2e = Mg$	−2.37

important to note, however, that these *equilibrium* potentials are actually thermodynamic quantities and have little relevance to the potentials of metals in solutions encountered in service, in which the potential of importance is the *corrosion potential.*

The equilibrium potentials provide information on the spontaneity of a reaction and how far it will proceed before equilibrium is achieved, but provide no information on the rate of the reaction. Thus the reduced form, the metal, of all equilibria below the hydrogen equilibria is oxidised by hydrogen ions at unit activity, and the tendency of the reaction to proceed in this direction is given by the e.m.f. of the cell. Algebraic summation of the relevant half-reactions and potentials provides information on the corrosion of metals in acid. For example

$$Fe + 2H^+ \rightarrow Fe^{2+} + H_2 \qquad E^{\ominus}_{cell} = 0.00 - (-0.44) = 0.44\ V$$

$$Zn + 2H^+ \rightarrow Zn^{2+} + H_2 \qquad E^{\ominus}_{cell} = 0.00 - (-0.76) = 0.76\ V$$

and the positive value of $E^{\ominus}_{cell}$ shows that both reactions proceed spontaneously in the direction in which they are written and that the tendency of zinc to react with hydrogen ions at $a_{H+} = 1$ is greater than that of iron. It can be calculated from $E^{\ominus}_{cell}$ that at equilibrium the pressure of hydrogen gas is $\approx 10^{20}\ N/m^2$ for the iron reaction and $\approx 10^{31}\ N/m^2$ for the zinc reaction.

In the case of the equilibria whose standard electrode potentials are more positive than 0.00 V the spontaneous direction of the reaction is in the reverse direction, and the stable form of the metal is the metal cation. Thus in the case of silver the reaction proceeds spontaneously in the direction

$$2Ag^+ + H_2 \rightarrow 2H^+ + 2Ag \qquad E^{\ominus}_{cell} = 0.79\ V$$

and the pressure of hydrogen at equilibrium is $10^{-22}\ N/m^2$. Thus the metals Au, Ag and Cu may be regarded as being stable in a reducing acid in which the sole oxidising species is the hydrogen ion.

Again it must be emphasised that although $E^{\ominus}_{cell}$ provides a measure of the tendency of the reaction to proceed, it provides no information on rates; for example, pure zinc corrodes more slowly in a reducing acid such as sulphuric than pure iron. Species other than the hydrogen ion can act as oxidants, and this applies particularly to dissolved oxygen, which is invariably present in aqueous environments in contact with the atmosphere. *Table 1.3* gives the reversible potentials of the two equilibria at different pH values, and it can be seen that dissolved oxygen has a more positive potential than the hydrogen ion. This means that the noble metals Cu and Ag will not corrode in a reducing acid in the absence of oxygen, but will do so in its presence.

The standard electrode potentials also indicate the relative tenden-

Table 1.3 REVERSIBLE POTENTIALS OF THE $H^+/\frac{1}{2}H_2$ AND $O_2/2OH^-$ EQUILIBRIA AT DIFFERENT pH VALUES AT 25°C($p_{H_2} = p_{O_2} \approx 10^5$ N/m²)

Equilibrium	*Activity*	pH	*E* (V)
$2H^+ + 2e = H_2$	$a_{H^+} = 1$	0	0.00
$E_r = 0.0 - 0.059\,\text{pH} - 0.059 \log p_{H_2}$	$a_{H^+} = 10^{-7}$	7	−0.414
	$a_{H^+} = 10^{-14}$	14	−0.828
$\frac{1}{2}O_2 + H_2O + 2e = 2OH^-$	$a_{OH^-} = 1$	14	0.401
$E_r = 1.23 - 0.059\,\text{pH} + 0.015 \log p_{O_2}$	$a_{OH^-} = 10^-$	7	0.815
	$a_{OH^-} = 10^{-14}$	0	1.230

cies of metal cations in aqueous solutions to be reduced at a cathode. Thus silver ions are more readily reduced than cupric ions, a fact that is utilised in the electrorefining of silver using a silver nitrate bath that deposits pure cathode silver even though it contains a high concentration of copper ions. Metal ions above the hydrogen equilibrium are more readily reduced than the hydrogen ion, and deposit at 100 per cent efficiency; in fact the cathodic reduction of Ag^+ and Cu^{2+} ions are utilised in the silver and copper coulometers to measure the passage of charge.

In the case of the metals that are more negative than the hydrogen equilibrium it might be predicted from the e.m.f. series that the reduction of hydrogen ions to hydrogen gas would be the preferred process, and that the more negative the equilibrium the more difficult would it be to obtain a high cathode efficiency. Such is not the case, and it is significant that a metal as negative as zinc ($E^\ominus_{Zn^{2+}/Zn} = -0.76$ V) can be plated from an acid sulphate bath at approximately 95 per cent efficiency, whereas Cr attains an efficiency of only 10–15 per cent when plated from a $CrO_3 + H_2SO_4$ bath. Even manganese ($E^\ominus_{Mn^{2+}/Mn} = -1.18$ V) can be deposited from an aqueous solution, but with more negative metals this is no longer possible and with a metal such as aluminium, hydrogen evolution is the sole cathodic reaction; for this reason aluminium can be deposited only from non-aqueous organic solutions or from fused salts.

Table 1.2 shows that the equilibrium e.m.f. of the Daniell cell, which must be determined in such a way that no current is drawn from the cell so that the potentials remain at their equilibrium values, is given by

$$Cu^{2+}(\text{aq.}) + 2e = Cu \qquad E^\ominus_{Cu^{2+}/Cu} = 0.34 \text{ V}$$

$$Zn = Zn^{2+}(\text{aq.}) + 2e \qquad E^\ominus_{Zn^{2+}/Zn} = -(-0.76) \text{ V}$$

By addition:

$$Cu^{2+}(\text{aq.}) + Zn = Zn^{2+}(\text{aq.}) + Cu \quad E^\ominus_{cell} = 1.1 \text{ V}$$

(Note that since the Zn/Zn^{2+}(aq.) equilibrium is written as an oxidation the sign must be reversed.)

This e.m.f., which is the maximum possible, is produced by the cell only when the rate of the cell reaction is practically zero. It will be seen that when the rate of the reaction is finite the e.m.f. decreases, and that the e.m.f. attains a minimum when the rate of reaction is at a maximum value, i.e. when the electrodes of the cell are short-circuited.

Polarisation

When a M^{z+}/M (aq.) system such as Cu^{2+} (aq.) $= 2e \rightleftharpoons$ Cu is at equilibrium the rates of the anodic reaction (Cu $\rightarrow$ Cu^{2+}(aq.)) and the cathodic reaction (Cu^{2+}(aq.) $\rightarrow$ Cu) are equal; there is no *net* passage of charge and no current will flow in an external circuit in either direction. The potential across the metal/solution interface is then the equilibrium potential $E_{eq.}$, which will conform with the Nernst equation provided the system is reversible; electrodes such as Cu^{2+}/Cu, $Hg_2{}^{2+}/Hg$, Zn^{2+}/Zn are of this type, whilst irreversible electrodes such as Ni^{2+}/Ni, Fe^{2+}/Fe, Al^{3+}/Al, etc. do not conform to the Nernst equation.

If the potential is displaced from its equilibrium value the electrode is said to be *polarised* and a net anodic or cathodic current will flow, depending upon whether the *polarised potential* E_p is more positive or more negative than E_{eq}, respectively. This displacement of the potential from its equilibrium value is called the *overpotential* η, and is defined by

$$\eta = E_p - E_{eq} \qquad (1.17)$$

Referring now to the Daniell cell, it has been shown that when no reaction occurs the electrodes are at equilibrium and the reversible e.m.f. of the cell reaction is 1.1 V, the maximum e.m.f. possible with this particular reaction. If now the resistance in the metallic circuit is decreased slightly, the electrodes become polarised, and consequently the e.m.f. becomes less than the maximum reversible value; a net current flows, the zinc being anodically oxidised to Zn^{2+}(aq.) and the Cu^{2+}(aq.) ions in solution being cathodically reduced to Cu metal. *Figure 1.6* shows how the polarised potential of the cathode $E_{p,c}$ becomes increasingly more negative while the polarised potential of the anode $E_{p,a}$ becomes increasingly more positive as the rate of the reaction, as given by the magnitude of the current I, is increased by decreasing the external resistance. Although the e.m.f. of the cell may be measured by a high impedance electrometer in the circuit, the

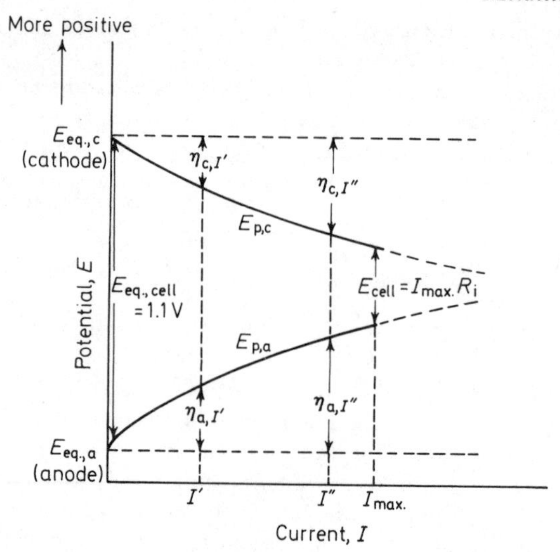

Figure 1.6 Potential–current relationships for the anodic and cathodic half-reactions constituting the Daniell cell reaction

potentials of the individual electrodes must be determined by a reference electrode and Luggin capillary, with the tip of the latter placed close to the surface of the electrode to minimise the IR drop (*Figure 1.7*).

These considerations may be expressed in a general form:

For a cathode $E_{p,c} < E_{eq,c}$ and $\eta_c < 0$, i.e. the cathode overpotential is always negative

For an anode $E_{p,a} > E_{eq,a}$ and $\eta_a > 0$, i.e. the anode overpotential is always positive

When the electrodes are short-circuited the current attains its maximum value and the e.m.f. its minimum value, and

$$E_{cell} = I_{max} R_i \tag{1.18}$$

where R_i is the resistance of the electrolyte solution.

In the Daniell cell the electrodes are well defined and physically separated so that the rate of charge transfer can be determined readily by means of an ammeter in the external circuit. The resistance of the electrolyte solution is significant, and under these circumstances $E_{p,c} > E_{p,a}$. It will be seen that a similar electrochemical mechanism applies to the corrosion of a metal in a solution, and that when the solution is highly conducting $E_{p,c} \approx E_{p,a} = E_{corr}$, where E_{corr} is the *corrosion potential.*

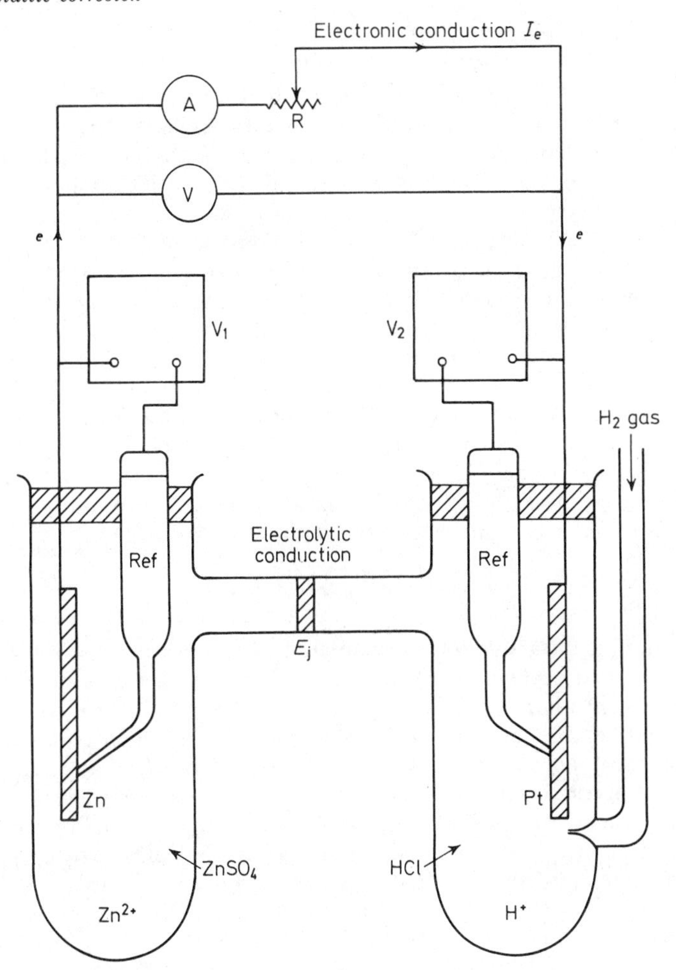

Figure 1.7 Method of determining the polarised potentials of the electrodes of an electrochemical cell by means of a reference electrode and Luggin capillary, which is placed close to the surface of the electrode to minimise the iR drop. The cell reaction is $Zn + 2H_3O^+ \rightarrow Zn^{2+} + H_2 + 2H_2O$, *and the cell may be represented as* $Zn|Zn^{2+}\ (aq.)|H_3O^+, H_2|Pt$, *in which the half-cell on the right-hand side is a reversible hydrogen electrode.*

Types of overpotential

Although a detailed exposition of overpotential is beyond the scope of this book it should be noted that three types of overpotential can be distinguished, i.e. *activation overpotential, concentration overpotential* (also known as transport overpotential) and *resistance overpotential.*

Activation overpotential

This is due to the fact that any electrode reaction that proceeds at a measurable rate requires an activation energy, which varies in magnitude with the system under consideration. Certain electrode reactions require only a small activation energy and this applies to the systems used as reference electrodes, such as Ag^+(aq.)/Ag, Hg_2^{2+}(aq.)/Hg, Cu^{2+}(aq.)/Cu, for which it is essential that the potential remains almost constant when small currents are drawn from or supplied to the electrode. These are referred to as *reversible* systems, and an ideal reversible electrode should remain unpolarised even when the reaction rate is very high; this is not possible, but the reference electrodes used in practice approximate to this ideal, provided the currents are small. At the other extreme are the *irreversible* electrodes; for example Ni in a solution of Ni^{2+}(aq.) polarises significantly even when the anodic or cathodic currents are very small. The relationship between activation overpotential and current density is given by the well known Tafel equation:

$$\eta_A = a \pm b \log i \qquad (1.19)$$

where η_A is the activation overpotential, i is the current density and a and b are constants that depend on the system under consideration, the concentration of ions, temperature, etc. For a cathode reaction (η is negative) the Tafel slope b is negative, and for an anodic reaction (η is positive) b is positive. It follows from equation 1.19 that when the rate of a reaction is controlled solely by the activation energy the activation overpotential η_A is linearly related to the logarithm of the current density; electrode reactions whose rate depends on the activation energy of electron transfer are referred to as *activation controlled.* Curves showing the η_A vs i and η vs log i relationships for anodic and cathodic reactions are shown in *Figures 1.8(a)* and *(b).*

Concentration overpotential

During a cathodic reaction of the type M^{Z+} (aq.) $\rightarrow M$ the concentration of ions at the surface of the electrode becomes less than that in the bulk solution; the converse applies to the anodic reaction. This results in a concentration gradient across the diffusion layer, which may be regarded as a thin static layer of solution adjacent to the electrode of thickness δ, and the replenishment of ions removed from the solution by the cathodic reaction occurs by diffusion through this layer. If the concentrations of the reacting ions in the bulk solution and at the electrode surface are c_b and c_e, respectively, the difference in potential, or overpotential, can be evaluated from the Nernst equation, and for a cathodic reaction:

$$E_e - E_b = \eta_T = \frac{RT}{zF} \ln \frac{c_e}{c_b} = \frac{0.059}{z} \log \frac{c_e}{c_b} \text{ at } 25°\text{C} \quad (1.20)$$

where E_e is the potential at the surface of the electrode and E_b its potential in the bulk solution. It is evident that since $c_e < c_b$, E_e is less than E_b and η_T is negative. As the rate of the reaction is increased c_e

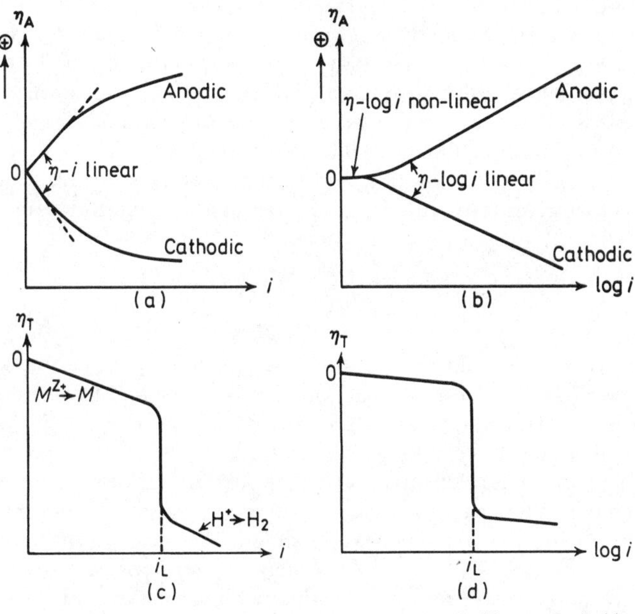

Figure 1.8 Overpotential–current density relationships: (a) η_A vs i; (b) η_A vs log i; (c) η_T vs i, showing the limiting current density i_L for $M^{z+} + ze \rightarrow M$ and the reduction of H^+ to H_2 gas at more negative potentials; (d) η_T vs log i

becomes smaller and smaller and eventually a rate is reached when c_e becomes zero, i.e. as soon as the metal ions reach the surface of the electrode they gain electrons and become metal atoms in the lattice of the cathode deposit. This limiting (or maximum) current density is given by

$$i_L = \frac{DzFc_b}{\delta} \quad (1.21)$$

where D is the diffusion coefficient ($\approx 10^{-6}$ cm^2 s^{-1} for most metal ions) and δ is the thickness of the diffusion layer, which varies from 0.05 cm for an unagitated solution to 0.001 cm for a highly agitated solution. Equation 1.21 shows that the limiting current density is increased by increasing the bulk concentration of the reacting ion, increasing the temperature and thus increasing the magnitude of the diffusion coefficient D, and by increasing the degree of agitation and thus decreasing the thickness of the diffusion layer δ.

The limiting current density defines the maximum current density for a specific electrode reaction, and any further increase in current can be achieved only by another electrode reaction; since hydrogen ions and water are invariably present this means that hydrogen evolution will accompany metal deposition at current densities above i_L for the latter reaction. It follows that in order to operate at a high current density and cathode efficiency a plating bath must contain a high concentration of the metal ions (aquo cations or anions) involved in the deposition of the metal, and that this will also be facilitated by agitating the plating solution and raising its temperature.

It is not possible to evaluate c_e directly, but by combining equations 1.20 and 1.21 it can be shown that

$$\eta_{T,c} = \frac{0.059}{z} \log \left(1 - \frac{i}{i_L}\right) \text{at } 25°\text{C} \tag{1.22}$$

and it is evident that the limiting current density is the important parameter in an electrode reaction that is controlled by diffusion of the reacting species. As an approximation $\log (1 - x) = x - \frac{1}{2}x^2 \ldots$ and neglecting all terms but the first, $\log (1 - x) \approx x$ and $\eta_{T,c} \approx 0.059/z\ (i/i_L)$, which means that at low current densities η is linearly related to i. When $i = i_L$, $\eta \rightarrow \infty$, i.e. η becomes infinitely negative, and no further increase in current is possible without an alternative reaction (see *Figure 1.8(c)* and *(d)*). In neutral solutions containing dissolved oxygen the latter provides the cathode reactant for the corrosion reaction, and the rate of corrosion is controlled by the rate of diffusion of oxygen to the metal surface; frequently, the rate of corrosion is equal to the limiting current density for oxygen diffusion.

Resistance overpotential

The resistance overpotential at an electrode is due to any iR drop resulting from the resistivity of the solution and/or the formation of solid films or deposits of corrosion products on the surface of the electrode, i.e.

$$\eta_R = IR_e + IR_f \tag{1.23}$$

where R_e is the resistance of the solution and R_f is the resistance of the film or deposited corrosion products. Where an electrochemical cell is involved this resistance overpotential reduces the magnitude of the maximum current produced by the short-circuited cell; for example, in the Daniell cell if the concentrations of Cu^{2+} and Zn^{2+} ions are maintained equal $I_{max.}$ decreases as the concentrations decrease, owing to the increase in the resistance of the solutions, although the reversible e.m.f. of the cell will be unchanged. In the cathodic protection of steel in sea water using sacrificial anodes, the current between the anode and the steel decreases with time owing to the formation of a calcareous scale (a mixture of $CaCO_3$ and $Mg(OH)_2$) on the surface of the steel. If aluminium is used as the sacrificial anode a protective film of $Al_2O_3 . H_2O$ may form on its surface and thus reduce the current to a value that is insufficient to protect the steel. It will be seen that in the case of discontinuous metal coatings on a metal substrate the geometry of the discontinuity and/or the formation of films or deposition of corrosion products can markedly decrease the galvanic current flowing between the two metals.

Corrosion cells

The Daniell cell has been used to exemplify an electrochemical cell in which a metal (Zn) is anodically oxidised to metal ions while a species in solution (Cu^{2+}(aq.)) is cathodically reduced to metal. It is evident from this cell that the rates and extents of the anodic and cathodic reactions must be equivalent, since they are dependent on the rate of transfer of charge (electrons) through the metallic part of the circuit. In this cell the two electrodes are physically separable and the rate of charge transfer can be determined readily by means of an ammeter in the circuit.

A similar mechanism applies to the corrosion of a single metal, but with the fundamental difference that the electrodes constituting the corrosion cell are not always identifiable. Thus in the *uniform* corrosion of a metal it is not possible to distinguish the anodic and cathodic sites, since they are of atomic dimensions and are constantly interchanging, i.e. at one instant of time a metal atom is sustaining a cathodic reaction and at another it is being removed anodically as a hydrated metal ion, the metal corroding so uniformly that the anodic and cathodic sites are indistinguishable. Thus corrosion is characterised by the fact that a metal is the reactant and also provides the means of transporting electrical charge from one part of its surface to another in the same way as the external metallic path in the Daniell cell.

However, corrosion is not always uniform and when attack is localised, although the anodic and cathodic areas may be distinguished visually, it is not possible to determine the rate of charge transfer by inserting an ammeter in the circuit. Bimetallic corrosion is an exception to this rule; for example it would be possible to study the effect of copper on the corrosion of zinc in oxygenated sodium chloride solution by coupling the two metals together through a zero-resistance ammeter and measuring I_{galv}, the galvanic current flowing from the zinc to the copper. Although this cell would be similar to the Daniell cell it is important to note that the cathodic reaction would be the reduction of dissolved oxygen to hydroxyl ions and not the reduction of cupric ions to copper.

Cathodic and anodic reactions in corrosion

It is evident from the discussion of the Daniell cell that for corrosion to take place the metal must be unstable in the solution under consideration, i.e. an electron acceptor must be present in solution with a higher redox potential than the M^{z+}/M system. Although Cu^{2+} ions act as cathode reactants in the Daniell cell, this is rare in practical cases of corrosion (certain types of corrosion of copper alloys are exceptions) and the two most common cathodic reactants in natural environments are the hydrated proton H_3O^+ (or the water molecule), and dissolved oxygen, which is invariably present when the aqueous environment is in contact with the atmosphere.

In the *hydrogen-evolution reaction* H_3O^+ ions or H_2O molecules are reduced to H_2 gas:

$$H_3O^+ + e \rightarrow \tfrac{1}{2}H_2 + H_2O \tag{1.24}$$

$$H_2O + e \rightarrow \tfrac{1}{2}H_2 + OH^- \tag{1.25}$$

In the *oxygen-reduction reaction* dissolved O_2 is reduced to OH^- ions:

$$O_2 + H_2O + e \rightarrow 2OH^- \tag{1.26}$$

It should be noted that these reactions are not mutually exclusive and that the type and number of reactions that occur will depend on the system under consideration. Thus in the case of Zn in oxygen-free acid reaction 1.24 is the sole reaction, whereas both reactions 1.24 and 1.26 are significant if the acid is oxygenated. Copper cannot be oxidised by H_3O^+ ions and remains unattacked in an oxygen-free reducing acid (see *Tables 1.2* and *1.3*) but corrodes if oxygen is present, reaction 1.26 being the sole cathodic reaction. Magnesium, a very electronegative metal ($E^{\ominus}_{Mg^{2+}/Mg} = -2.1$ V), corrodes in oxygen-free neutral sodium chloride with hydrogen evolution, whereas iron in these circumstances remains unattacked. On the other hand, in many corrosion reactions involving oxygen-containing solutions the hydrogen-evolution reaction and oxygen-reduction reaction occur simultaneously. The relative roles played by oxygen, the hydrated proton and the water molecule in corrosion are highly complex and depend upon such factors as the nature of the metal, the pH of the solution, the concentration of dissolved oxygen, temperature, complex formation, etc. The rate of the hydrogen-evolution reaction is normally activation controlled and depends very markedly on the nature of the electrode, although the pH of the solution, temperature, etc., also have an effect. Thus the relationship between overpotential and current density conforms with the Tafel equation (equation 1.19) with a and b varying with the nature of the metal and the composition of the solution. At high current densities transport becomes significant and the linear relationship between η and $\log i$ no longer prevails. The converse applies to the oxygen-reduction reaction and, although activation control is significant at low current densities, diffusion becomes more significant at higher current densities and the corrosion rate then corresponds with the limiting current density. It should be noted that, unlike activation overpotential, concentration overpotential is not dependent on the nature of the electrode, although the presence of films and corrosion products that impede electron transfer at the cathodic sites will markedly affect the rate.

In the Daniell cell the zinc is anodically oxidised to Zn^{2+}(aq.), but although this type of reaction is extremely important it is by no means the only one possible, and in general it is possible to distinguish three distinct types of anodic reactions resulting in the formation of metal cations, metal anions and solid metal compounds (not necessarily oxides, hydroxides or hydrated oxides):

$$M \rightarrow M^{z+}(\text{aq.}) + ze \quad \text{metal cation} \quad (1.27)$$

$$M + zH_2O \rightarrow MO_z^{z-} + 2zH^+ + ze \quad \text{metal anion} \quad (1.28)$$

$$M + zH_2O \rightarrow M(OH)_z + zH^+ + ze \quad \text{metal oxide (or solid compound)} \quad (1.29)$$

Metal cations and anions are mobile and are able to diffuse away from the surface of the metal, so continued dissolution is unimpeded, although the rate may be affected by concentration overpotential. However, if the anodic reaction results in the formation of a solid oxide or other solid compounds (e.g. the formation of Al_2O_3 on Al in H_2O) this may result in the formation of a barrier between the metal and the aqueous environment so that further reaction must now take place via the intervening layer of compound. The ability of the solid compound to protect the metal will, of course, depend upon its solubility in the environment, its adherence to the surface of the metal, the cohesion of its crystals, etc., and different metal/environment systems give rise to layers of solid compounds that differ in the degree of protection that they afford to the metal. Metals like Ni, Cr, Al, Ti and the stainless steels have the ability to form thin invisible films of oxide (1–3 nm thick) in a number of environments, and although these metals are electrochemically active the films have a very marked effect on the rate of reaction. This ability of a metal to form a protective film is known as *passivity,* and the passivation of a metal is one of the most important methods of corrosion control. Some metals are passive in a variety of environments while others are passive only under very specific environmental conditions; in this connection tantalum and iron represent two extremes — tantalum is passive in most acids including hydrochloric acid, whereas iron is passive only in fuming nitric acid.

Since mild steel is the most important substrate for metallic coatings it is appropriate to consider its corrosion in aqueous solutions. In oxygenated sea water the cathode reaction is oxygen reduction, and the anodic reaction proceeds by the following steps:

$$Fe \rightarrow Fe^{2+} + 2e$$

$$Fe^{2+} + OH^- \rightarrow \underset{\text{white-green precipitate}}{Fe(OH)_2}$$

$$4Fe(OH)_2 + O_2 \rightarrow \underset{\text{red-brown rust}}{2Fe_2O_3.H_2O} + 2H_2O$$

The primary step is the formation of ferrous ions, which are mobile and can diffuse and migrate away from the surface of the metal, and since the hydroxyl anions formed from the cathodic reduction of dissolved oxygen will move in the opposite direction the formation of ferrous hydroxide will occur at some intermediate position between the anodic and cathodic sites. This will be followed by the chemical oxidation of ferrous hydroxide to hydrated ferric oxide or rust by the oxygen present in the water. The important point to note is that since the rust is formed away from the surface of the metal it cannot have any effect on the rate of corrosion. *Figure 1.9* shows a hypothetical cell that illustrates the corrosion of steel in oxygenated sea water. *Figure 1.10* shows the reactions when a droplet of NaCl solution is placed on the surface of a steel sheet, in which the geometry of the system is such that the periphery of the droplet, through which atmos-

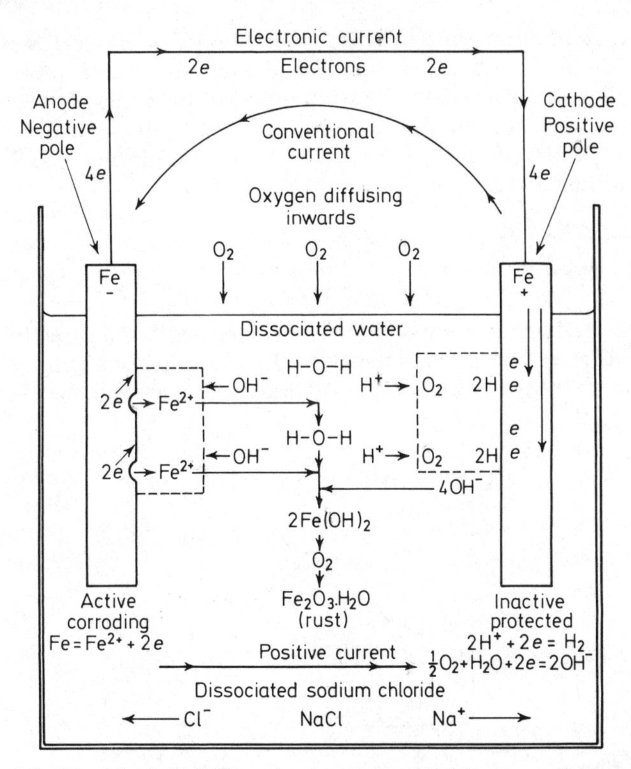

Figure 1.9 Electrochemical cell illustrating the corrosion of iron in oxygenated sodium chloride solution, in which the inseparable anodic and cathodic sites arising from the corrosion reaction have been represented by well defined separable electrodes (From T. Howard Rogers, Marine Corrosion, George Newnes, 1968)

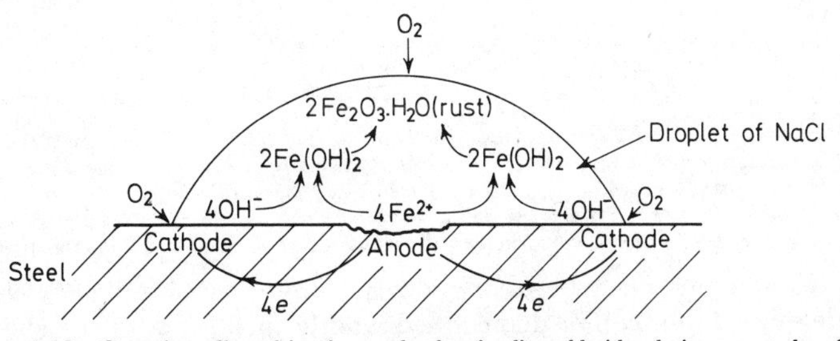

Figure 1.10 Corrosion cell resulting from a droplet of sodium chloride solution on a steel surface. The geometry of the droplet results in a situation where easy access of oxygen to the periphery of the steel as compared to its interior results in the former becoming the cathode and the latter the anode of the corrosion cell

pheric oxygen can diffuse rapidly to the surface of the steel, becomes the cathodic area while the interior becomes anodic. However, it is possible to make the iron passive in water by adding a corrosion inhibitor such as sodium chromate, which results in the formation of a protective film of Fe_2O_3 at weak areas in the air-formed oxide film, thus stifling attack.

Evans diagrams

The mutual polarisation of the inseparable anodic and cathodic sites at the surface of a corroding metal is in many respects analogous to the mutual polarisation of the well defined and separable electrodes of

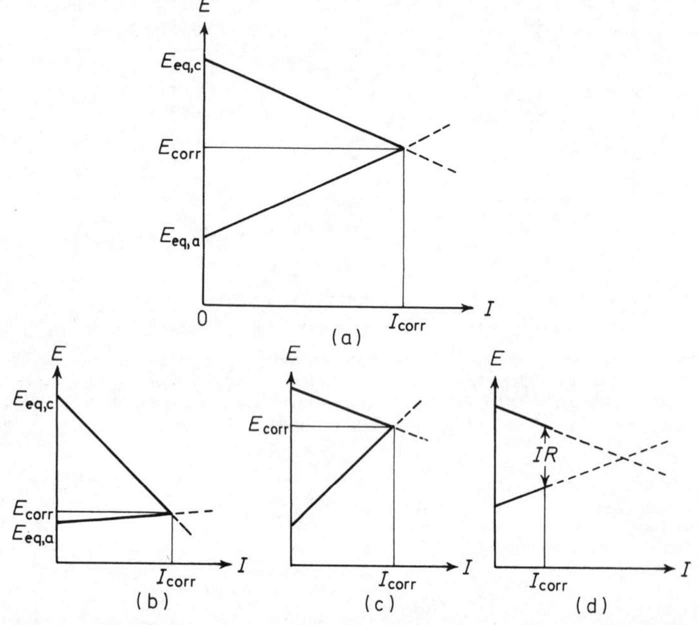

Figure 1.11 Evans E–I diagrams to illustrate the corrosion of metals: (a) E–I diagram showing how the mutual polarisation of the anodic and cathodic reaction defines the corrosion potential E_{corr} and the corrosion current I_{corr}; (b) cathodic control, in which the pronounced polarisation of the cathodic curve determines the rate of corrosion; (c) anodic control, in which the pronounced polarisation of the anodic curve determines the rate of corrosion; (d) resistance control, in which IR drops limit I_{corr}

the Daniell cell, but whereas the degree of polarisation of each electrode of the latter can be determined readily such is not the case with the former. Diagrams illustrating corrosion of metals by means of the E vs I curves for the anodic and cathodic reactions are known as *Evans diagrams*, and *Figure 1.11(a)* shows how the intersection of these two curves (drawn as straight lines) defines the corrosion potential E_{corr} and the corrosion rate I_{corr}. At $I = 0$, the electrodes are at equilibrium and $E_{eq,a}$ and $E_{eq,c}$ are the equilibrium potentials of the anodic and cathodic reactions, respectively; however, it should be noted that the actual potential difference at the metal/solution interface is E_{corr} and not $E_{eq,a}$. Although E_{corr} can be measured by means of a reference electrode, this is not the case with I_{corr}, since it is obviously impossible to insert an ammeter in the circuit. However, I_{corr} can be evaluated indirectly from the weight loss of the metal/unit time and from Faraday's law. *Figures 1.11 (b), (c)* and *(d)* show how the kinetics of the reactions control the corrosion rate by cathodic, anodic and mixed control, respectively.

Figure 1.12 illustrates a more fundamental approach in which the E–i curves have been replaced by E–log i curves, since the latter are applicable when the anodic and cathodic reactions are activation controlled.

When oxygen is the cathode reactant the rate is controlled by i_L, the limiting current density, and *Figure 1.13* illustrates how the corrosion

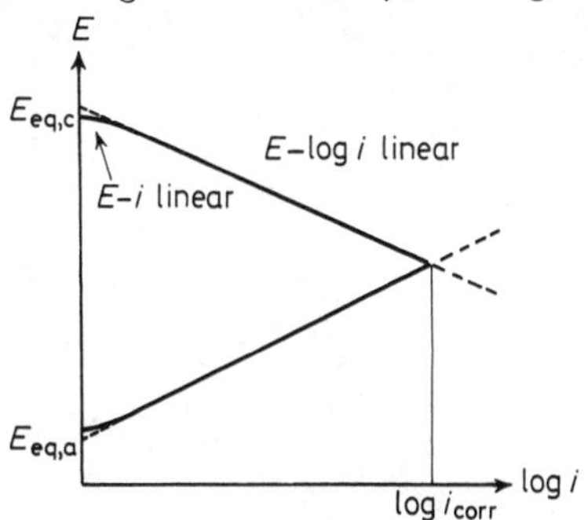

Figure 1.12 E–log i curve for a metal in which both the cathodic and anodic reactions are activation controlled so that at overpotential >0.05 V the curves conform to the Tafel equation

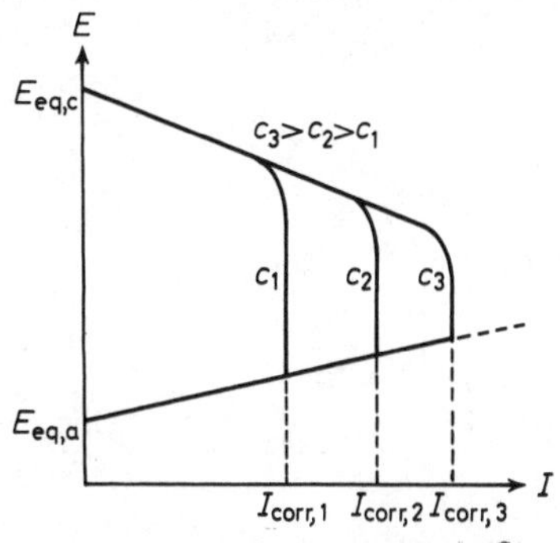

Figure 1.13 E–I diagrams illustrating how the corrosion rate of steel in oxygenated water increases with the concentration of dissolved oxygen; similar curves can be drawn to illustrate the effect of increase in velocity. Note that I_{corr} (or i_{corr}) is dependent on the limiting diffusion current I_L (or i_L)

rate of iron increases with concentration of oxygen in the water. It can be seen that when the rate of a corrosion reaction is controlled by cathodic reduction of dissolved oxygen $I_{corr} = I_L$, where I_L is the limiting current (or limiting current density) for oxygen reduction.

Dissimilar metals in contact

These considerations emphasise the fact that in most metal/solution systems the potential of the metal is a mixed or corrosion potential resulting from the mutual polarisation of the anodic and cathodic reactions. The standard electrode potentials are therefore of little practical significance, and it is unfortunate that they are referred to when considering various corrosion phenomena; this applies particularly to their misuse in attempting to predict the effects produced when two dissimilar metals are in metallic contact.

When two dissimilar metals M_A and M_C are in metallic contact their corrosion rates usually differ from those of the metals when uncoupled. This depends on their respective corrosion potentials (not the standard electrode potentials), and if the corrosion potential of metal M_A is more negative than that of M_C electrons will be transferred from M_A to M_C with a consequent increase in the potential (more positive) of M_A and a decrease in the potential of M_C (more negative). This results in an increase in the anodic current and a decrease in the cathodic current of M_A, which will corrode more rapidly, while the converse applies to M_C, whose corrosion rate will decrease (*Figure 1.14*). Thus if the metals are coupled through a vari-

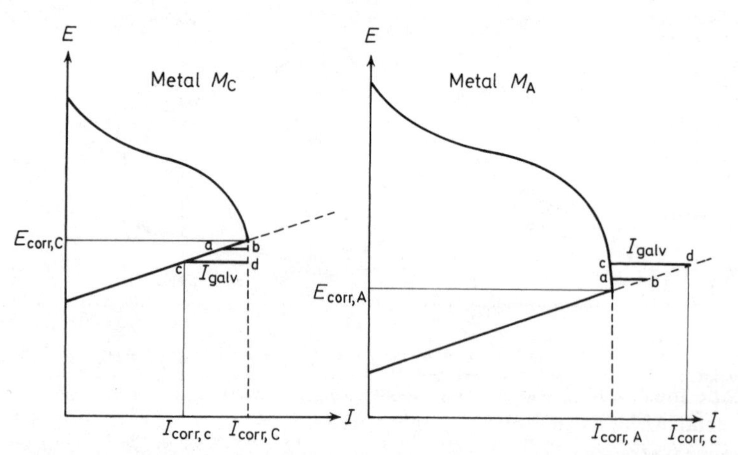

Figure 1.14 Effect of coupling two dissimilar metals M_A and M_C, where M_A has a more negative corrosion potential than M_C so that electron transfer occurs in the direction $M_A \rightarrow M_C$. This results in the potential of M_A becoming more positive, with consequent increase in the corrosion rate from $I_{corr,A}$ to $I_{corr,C}$ whilst the potential of M_C becomes more negative with consequent decrease in the corrosion rate from $I_{corr,C}$ to $I_{corr,c}$. The transfer of charge, which can be determined by means of a zero-resistance ammeter placed between M_A and M_C, is the galvanic current I_{galv}, i.e. ab, cd

able resistance the rate of charge transfer (the current *I*) from M_A to M_C may be equal to $I =$ ab, and the potentials will not be equal. If the two metals are small in area and are short-circuited (in direct contact), and if the solution is of high conductivity, the potentials will become equal by a transfer of charge equal in magnitude to cd. The current transferred between M_A and M_C is called the galvanic current, I_{galv}, and gives a measure of the enhanced corrosion that the former will suffer when coupled to M_C. This illustrates the danger of coupling dissimilar metals. For example it would be highly detrimental to place brass in contact with a zinc-base diecasting since the former would stimulate the rate of attack of the latter. On the other hand, metals like zinc, magnesium and aluminium are deliberately sacrificial in the protection of steel structures in natural waters such as ships, jetties, drilling platforms, etc., a method of corrosion control known as *cathodic protection.* Furthermore, coatings of zinc on steel have a similar effect at discontinuities, although there are important differences between the sacrifical action of a zinc coating on steel (with the metals in direct contact) and the cathodic protection of a steel structure by a zinc anode that is located away from the structure (with electrical contact being made by a suitable connector). Although the individual corrosion potentials of two dissimilar metals indicate the direction of transfer of charge when the two metals are coupled together, they do not provide information on the intensity of attack (rate per unit area or current density) on the more negative metal. This depends on a number of factors such as:

(a) the nature of the environment, which determines whether oxygen reduction or hydrogen evolution will be the dominant cathodic reaction;
(b) the rate of the cathodic reaction on the more positive metal, M_C;
(c) the formation of protective films on both metals, which will impede electron transfer at M_C and will increase the polarisation of M_A;
(d) the relative effective areas (not necessarily the superficial areas) of M_A and M_C;
(e) the conductivity of the solution.

The most dangerous situation is when the area of the more positive metal M_C is large and that of the more negative metal M_A is small, since under these circumstances I_{galv} (the current flowing from M_A to M_C) is large and results in a large anodic current density at M_A ($i_A = I_{galv}/S_A$). In neutral solutions the cathode reactant is dissolved oxygen, and I_{galv} (the current flowing from M_A to M_C) increases with the rate at which the oxygen is brought to the surface of M_C (the rate increases with concentration of oxygen and velocity); if this is constant the rate is approximately proportional to the area of M_C. In acid solutions similar area relationships apply, but the nature of the metal M_C is also of importance since the hydrogen evolution reaction is then the predominant cathodic reaction and the rate of discharge at a given potential is dependent on the Tafel constants, a and b.

Films on the surface of the cathodic metal that impede the transfer of electrons have a significant effect on the rate, and it is for this reason that the noble metals Ag, Cu, Au, Pt, which remain unfilmed, are more detrimental to the more anodic metal than metals such as Fe and Pb in neutral solutions in which dissolved oxygen is the cathode reactant. These noble metals have a similar effect in acid solutions, but the reason in this case is their ability to catalyse the hydrogen-evolution reaction, which occurs at low overpotentials on these metals. The conductivity of the solution determines the areas of the two metals M_A and M_C that are effective in bimetallic corrosion, and it must be emphasised that these areas correspond with the superficial areas only when the solution is highly conductive and the areas are relatively small. In solutions of low conductivity the effective areas are largely confined to the areas adjacent to the interface between the two metals, and although under these circumstances the anodic area is small attack is not intense owing to the correspondingly small area of the cathode. The effect of the resistivity of the electrolyte, geometry of the system and formation of films will be considered in more detail in relation to metal coatings.

Potentials of metals in practice

It follows from previous considerations that the potential of a metal in a practical environment is its corrosion potential, which is determined by the nature of the anodic and cathodic reactions constituting the overall corrosion reaction. Thus in contrast to the standard electrode potential, which is a constant for a given equilibrium, the corrosion potential varies with the nature of the environment, temperature, velocity, etc. There are a variety of tables that give information on the potentials of metals in different environments, but in view of the importance of sea water as a corrosive environment this has been given the greatest attention and the results have been embodied in the so-called *Galvanic Series.* This table (*Table 1.4*) does not give actual values of the potential (which vary with the composition of the sea water, its degree of aeration, temperature and velocity) but arranges the metals in the order of their typical corrosion potentials in this environment, with the most noble (positive) at the top and the most active at the bottom; the further apart the two metals in the series, the greater the galvanic effect when they are coupled. For comparison, *Table 1.5* gives values determined in aerated moving sea water determined by means of a saturated calomel electrode (SCE) and expressed with reference to this electrode.

It should be noted that neither the distance apart of the two metals in the galvanic series nor their actual difference in potential provides information on the magnitude of the galvanic current, since this

Table 1.4 GALVANIC SERIES OF SOME COMMERCIAL METALS AND ALLOYS IN SEA WATER*

↑	Platinum
	Gold
Noble or	Graphite
cathodic	Titanium
	Silver
	Chlorimet *3* (62Ni–18Cr–18Mo)
	Hastelloy *C* (62Ni–17Cr–15Mo)
	18/8 Mo stainless steel (passive)
	18/8 stainless steel (passive)
	Chromium stainless steel 11–30% Cr (passive)
	Inconel (passive) (80Ni–13Cr–7Fe)
	Nickel (passive)
	Silver solder
	Monel (Ni–30Cu)
	Cupro-nickels (60–90Cu, 40–10Ni)
	Bronzes (Cu–Sn)
	Copper
	Brasses (Cu–Zn)
	Chlorimet *2* (66Ni–32Mo–1Fe)
	Hastelloy *B* (60Ni–30Mo–6Fe–1Mn)
	Inconel (active)
	Nickel (active)
	Tin
	Lead
	Lead-tin solders
	18/8 Mo stainless steel (active)
	18/8 stainless steel (active)
	Ni-Resist (high nickel cast iron)
	Chromium stainless steel, 13% Cr (active)
	Cast iron
	Steel or iron
	2024 aluminium (Al–4·5Cu–1·5Mg–0·6Mn)
Active or	Cadmium
anodic	Commercially pure aluminium (1100)
↓	Zinc
	Magnesium and magnesium alloys

*Data after Fontana, M. G., and Greene, N. D., *Corrosion Engineering*, McGraw Hill (1967).

depends on the kinetics of the cathodic and anodic reactions, the resistivity of the solution, film formation, relative effective areas of the two metals, etc. The galvanic current can of course be determined experimentally by direct measurement by means of a zero-resistance ammeter and a suitably designed bimetallic couple immersed in the environment under consideration. As a rough approximation the further apart the two metals are in the galvanic series, or the greater the e.m.f. of the couple, the greater the galvanic current, but there are many exceptions to this rule. Thus platinum and mercury have similar potentials in sea water (~ 0.0 V vs SHE), and although coupling platinum to magnesium (~ −1.0 V vs SHE) markedly increases the corrosion rate of the latter mercury has little effect. This is because magnesium in sea water corrodes with hydrogen evolution, and whereas platinum is a good catalyst for the hydrogen-evolution reaction mercury is not.

Table 1.5 POTENTIALS OF METALS IN AERATED MOVING SEA WATER (POTENTIALS ARE NEGATIVE TO THE SCE, E_{sce} = 0.246 V vs SHE)

Metal	*Potential* (V)
Magnesium	1·5
Zinc	1·03
Aluminium	0·79
Cadmium	0·7
Steel	0·61
Lead	0·5
Solder (50/50)	0·45
Tin	0·42
Naval brass	0·30
Copper	0·28
Aluminium brass	0·27
Gun metal	0·26
Cupro-nickel 90/10	0.26
Cupro-nickel 80/20	0·25
Cupro-nickel 70/30	0·25
Nickel	0·14
Silver	0·13
Titanium	0·10
Stainless steel 18/8 (passive)	0·08
Stainless steel 18/8 (active)	0·53

The new BSI Draft Commentary on corrosion at bimetallic contacts and its alleviation now being prepared includes detailed information in the form of a code (0, no additional corrosion; 1, slight additional corrosion; 2, fairly severe additional corrosion; 3, very severe additional corrosion) which indicates how the metal under consideration is affected when coupled to other metals during exposure to the atmosphere or immersion in natural waters. However, the Draft Document does not include information on the behaviour of bimetallic couples in chemical solutions or in foodstuffs, and under these conditions it is necessary to carry out corrosion tests.

The corrosion potential is a variable that is influenced by the specific environmental conditions prevailing, and it is of interest therefore to consider how these potentials differ from those given in the e.m.f. series of metals under different environmental conditions. Since mild steel is the most important substrate for metallic coatings it is appropriate to take the standard electrode potential of the Fe^{2+}/Fe equilibrium as the reference potential ($E^{\ominus}_{Fe^{2+}/Fe} = -0.44$ V), and on this basis metals used as coatings can be classified as follows:

(a) more positive than iron: Pt, Au, Ag, Cu, Sn, Ni, Cd
(b) more negative than iron: Cr, Zn, Al, Mg

It would appear that whereas the former metals should stimulate attack on steel at a discontinuity, the latter should cathodically protect the steel and in doing so should suffer enhanced corrosion. However, since the corrosion potentials seldom correspond with the standard electrode potentials this classification is often misleading.

Cathodic metals

The noble metals behave in practice according to their position in the e.m.f. series, but reference to the galvanic series (*Table 1.4*) shows that although copper is a noble metal ($E^{\ominus}_{Cu^{2+}/Cu} = 0.34$ V) its corrosion potential in sea water is more negative than the high nickel alloys (e.g. Hastelloy) and the stainless steels, provided these alloys are in the 'passive' condition. On the other hand, the stainless steels when in the 'active' condition have potentials similar to that of mild steel. This means that whereas passive 18Cr–8Ni stainless steel stimulates attack on copper and copper alloys, the converse applies when it is in the active condition. *Figure 1.15* shows the potentiostatically-determined anodic curve ABCD for a metal/environment system that shows an active-passive transition at B, and it can be seen that as the potential is made more positive the current density increases in the active region AB until a critical value is attained (the critical current-density i_{crit}) when the rate decreases suddenly owing to the formation of a protective film of oxide on the metal surface. The metal is then said to be passive, and its rate of corrosion, which is controlled by the oxide film, is significantly less than when in the active condition. As can be seen from *Figure 1.15* passivity can be achieved also by the redox potential of the solution and the kinetics of the cathodic reaction. Curve IJK represents the cathodic reduction of H^+ ions when the metal is corroding actively in a reducing acid, and it can be seen that the corrosion rate and corrosion potential are given by the intersection of this curve and the anodic curve at J. In a solution of high redox potential, such as may be achieved by saturating the reducing acid with oxygen or by adding an oxidising species such as nitric acid, the cathodic curve FGH intersects the anodic curve in the passive region at G, with a consequent decrease in the corrosion rate. It is evident that the metal when corroding actively has a more negative potential than when in the passive state, and this explains the two different positions of active/passive metals and alloys in the galvanic series. Thus in oxygenated waters, including condensed moisture from the atmosphere, metals like Ni and Cr are passive and have more

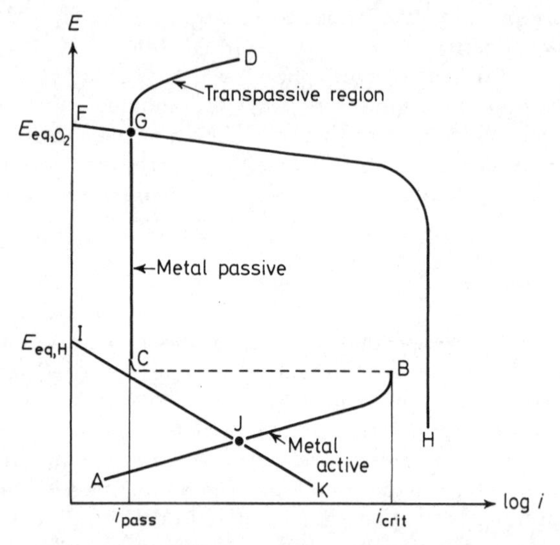

Figure 1.15 Potentiostatically-determined anodic curve ABCD for a metal/environment system that shows an active-passive transition at B with a consequent decrease in the corrosion rate from i_{crit} to i_{pass}. In a reducing acid of low redox potential the cathodic curve intersects the anodic curve in the active region at J, whereas in an oxidising acid or in reducing acid containing an oxidising species such as oxygen the cathodic curve FH intersects the anodic curve in the passive region at G

positive potentials than steel; the galvanic relationship between these metals, which form the composite steel/Ni deposit/Cr deposit, is discussed in more detail subsequently.

Although nickel corrodes in the active region with the formation of Ni^{2+} ions, this reaction requires a much higher activation overpotential than the anodic dissolution of reversible metals such as copper and zinc. However, in the case of nickel the overpotential is decreased significantly when sulphide ions are present in the solution. Advantage is taken of this phenomenon in the manufacture of electrolytic nickel anodes for use in the electroplating of nickel, which are manufactured by deposition from a nickel bath containing an organic sulphur compound so that a controlled amount of sulphur (0.02 per cent) is incorporated in the deposit. These anodes corrode anodically much more smoothly and uniformly than sulphur-free anodes, and at a more negative corrosion potential. A similar introduction of sulphur occurs in electroplating of bright-nickel deposits from baths containing organic sulphur compounds, which are used as levellers and brighteners. The sulphur-containing deposits are more electrochemically active, and hence more negative at a given corrosion rate, than dull-nickel plated from a plain Watts bath. As will be seen advantage is taken of this phenomenon in the protection of steel with a duplex nickel coating.

Anodic metals

The corrosion potential of a metal is frequently more negative than would be anticipated from the e.m.f. series; this applies to metals such as cadmium and tin, which under certain environmental conditions will sacrificially protect the steel substrate. Conversely, metals such as aluminium and zinc, which according to the e.m.f. series are significantly more negative than steel, may have corrosion potentials that make them cathodic to steel. This variable polarity depends, of course, upon the environmental conditions prevailing, and in certain systems a change in polarity results from only slight changes in the environment.

According to the e.m.f. series cadmium ($E^\ominus_{Cd^{2+}/Cd} = -0.403$ V) is 0.037 V more positive than steel, but it can be seen from the galvanic series that the positions are reversed. This can be explained by the anodic and cathodic *E–I* curves for the metals corroding in aerated waters, and it can be seen from *Figure 1.16* that the anodic curve for $Cd \rightarrow Cd^{2+}$ (Cd is a reversible metal) shows far less polarisation than that for $Fe \rightarrow Fe^{2+}$. Consequently corroding cadmium is slightly more negative than corroding iron, and sacrificially protects it.

It would appear from both the e.m.f. series and the galvanic series that tin is cathodic to steel, and this applies in solutions of inorganic salts or in natural water (including condensed moisture from the atmosphere). However, there are a number of exceptions: tin is anodic to steel in solutions of certain organic acids (citric, tartaric, oxalic, malic) and their salts, in fruit juices containing these acids, in meat and meat derivatives and in alkaline solutions. In this connection it should be noted that tin in Sn^{2+} ions forms a reversible electrode, and that tin has a strong tendency to form complexes with organic acids, with a consequent decrease in the activity of Sn^{2+} to a low value (see equation 1.14). Under these circumstances the potential of the tin will become far more negative than the standard electrode potential and will thus become anodic to steel.

It would appear from the e.m.f. series that both zinc and aluminium are negative to steel, and that aluminium is ≈ 1.0 V more negative than zinc. However, the potentials of both these metals are affected by the nature of the film formed on their surface, and this applies particularly to aluminium.

In most environments the corrosion potential of zinc is negative to steel, and this applies to sea water and to natural fresh water at ambient temperatures. However, in fresh waters, although zinc is anodic to steel below 60°C, a reversal in polarity occurs at this temperature and above: the zinc becomes cathodic to steel and stimulates attack at a discontinuity. This change has been attributed to the conversion of the zinc hydroxide film (a poor conductor of electrons) to a more conducting film of ZnO; a similar change with temperature

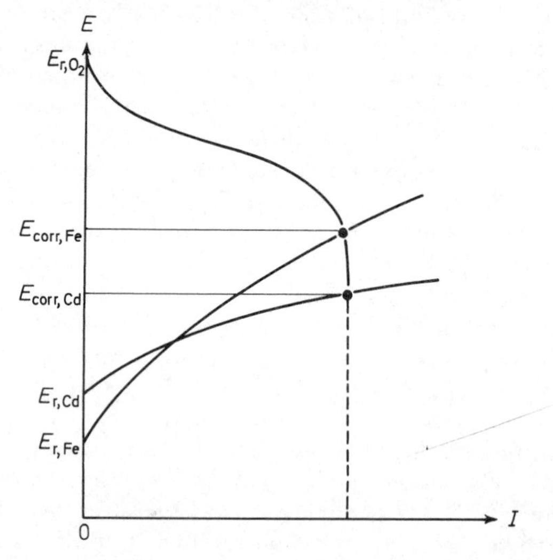

Figure 1.16 E–I curves illustrating how the relatively small polarisation of $Cd \rightarrow Cd^{2+}$ compared with the higher polarisation for $Fe \rightarrow Fe^{2+}$ results in a more negative corrosion potential of Cd in oxygenated water, so Cd will sacrificially protect a steel substrate at a discontinuity.

of water occurs in the case of the aluminium-iron couple; this too is considered to be due to a change in the nature of the oxide film. In sea water or chloride solutions, although the difference in potential between Fe and Zn decreases continuously with increase in temperature, zinc remains anodic to steel at all temperatures. It should be noted that zinc in sea water has a potential of −0.70 V, which is close to the reversible potential, and that it can be used as a reference electrode in this environment. The corrosion behaviour of aluminium is controlled by the presence of a film of oxide (in water at ambient temperatures the film is Bayerite, $Al_2O_3.3H_2O$), which provides the cathode for the anodic dissolution reaction that takes place at discontinuities in the film. Since the film is a poor conductor of electrons the cathodes are confined to areas of the film that are sufficiently thin to allow electron transfer. This results in a corrosion potential that is normally positive to zinc and (depending on circumstances) may be positive or negative to steel. The potential of aluminium in sea water is -0.55 V, i.e. only ≈ 0.10 V more negative to steel, but this difference in potential is sufficient to protect the latter sacrificially, and aluminium anodes are widely used for the cathodic protection of steel structures in sea water. In atmospheric environments, aluminium sometimes requires an induction period before its anodic behaviour becomes apparent. Thus sprayed aluminium coatings on steel when first exposed to the atmosphere show superficial rust stain due to the corrosion of the underlying steel exposed at pores, but after a short time these disappear owing to the penetration of the oxide film surrounding each aluminium particle, and the aluminium then sacrificially protects the steel and prevents it rusting. The oxide film on aluminium is made more electron-conducting if other ions enter the oxide lattice, and this applies particularly to copper ions; waters containing traces of dissolved copper can give rise to severe pitting of aluminium.

Effects of discontinuities in coatings

Previous considerations have been confined to the galvanic effects produced when two dissimilar bulk metals are in contact, and it is apparent that the metal with the more electropositive corrosion potential will be protected and will stimulate attack on the one that is less electropositive. Similar considerations apply to a discontinuous metal coating on a metal substrate, although the geometry of the discontinuity frequently has a controlling influence on the rate of attack. In this connection it must be emphasised again that the primary function of a coating is to provide a barrier that is more resistant to attack than the substrate, but since coatings are seldom continuous the substrate metal will be exposed to the environment and the galvanic relationship between the coating and the substrate will determine whether attack will be stimulated or stifled. On the basis of previous considerations it might be predicted that when the coating is anodic to the substrate the latter will be sacrificially protected, whereas attack will be stimulated when the coating is cathodic. However, this is an oversimplification of the situation, and in practice these effects are modified by the geometry of the discontinuity and by the nature of the environment.

Discontinuities in coatings include the fine and gross pores that arise from methods used for applying the coating and from defects in the substrate, stress cracks in certain electrodeposits such as chromium and rhodium, cut edges and damaged areas resulting from fabrication and maltreatment in service, etc. In general, porosity decreases with thickness of coating irrespective of method of application, and in the case of hot-dipped tin coatings on steel there is a logarithmic relationship between the number of pores/unit area and the coating thickness. Economics frequently determines coating thickness. The very thin electrodeposited tin coating on steel for the manufacture of tin cans is highly porous, but this is of little consequence since it is normally protected by a lacquer. On the other hand, the much thicker hot-dipped tin coatings used for food manufacture are practically non-porous.

Anodic coatings

Where corrosion-protection is of primary importance and appearance is of less importance, coatings of zinc, aluminium and cadmium are used to protect steel, and these metals have the advantage that they sacrificially protect the substrate at discontinuities under most environmental conditions. However, this results in the removal of the coating so that its function as barrier is lost. It follows that the magnitude of the galvanic current flowing between the coating and substrate should be only just sufficient to protect the latter.

It has been pointed out that the corrosion potential of a metal is a variable and is dependent on the environmental conditions prevailing, but in most environments the order of negative character is:

$$\mathrm{Sn} < \mathrm{Al} < \mathrm{Cd} < \mathrm{Zn}$$

The advantage of these metals is that the substrate is protected, which is important where thin sections of the latter are involved since a cathodic coating could lead to perforation; it is also important in avoiding the formation of unsightly corrosion products of the substrate.

Figure 1.17 (a) to *(c)* illustrates the action of an anodic coating in sacrificially protecting the substrate at a discontinuity with consequent consumption of the coating. If this process progresses too far the coating is no longer able to act as a barrier or to protect the substrate sacrificially at areas some distance from the coating; the 'throw' of the cathodic current of course depends on the corrosion potential of the coating and the conductivity of the solution within the discontinuity. A discontinuous magnesium coating on steel would fall into this category and in a highly conducting solution, such as sea water, its protective action (both as a barrier and as a sacrificial anode) would persist for only a short time.

Zinc is normally anodic to steel (E_{corr} = −0.7 V vs SHE in sea water) and sacrificially protects the latter in most environments. This protection persists even in the case of gross discontinuities, such as can be produced by bare cut edges; it has been demonstrated that steel remains protected in an industrial environment even when the coating is removed mechanically from the surface to give a strip of bare steel of 8–10 mm or more in width.

Figure 1.17 (g) illustrates the situation that prevails in practice

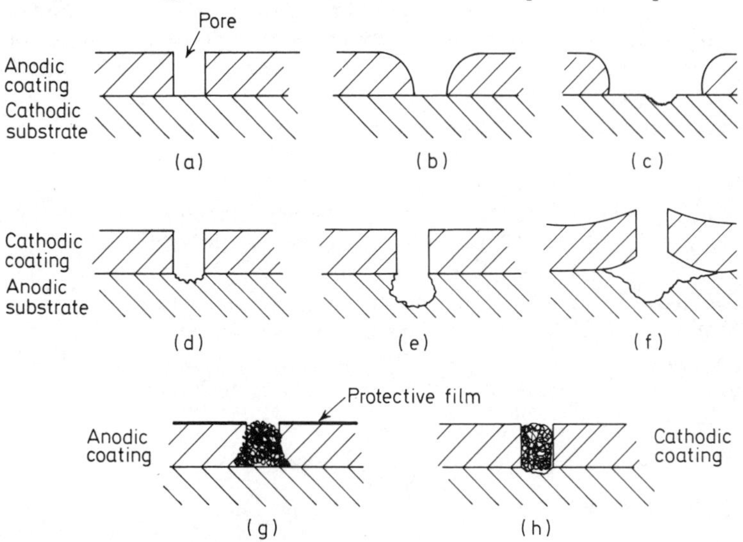

Figure 1.17 Diagrammatic representation of galvanic effects produced at a pore in a coating on a substrate; (a) – (c) anodic coating sacrificially protects cathodic substrate; (d) – (f) attack on anodic substrate is enhanced by cathodic coating, leading to pitting of substrate and exfoliation of the coating; (g) – (h) effect of corrosion plugs in plugging the pore, resulting in resistance control

when steel is exposed at a discontinuity in a zinc coating, and it can be seen that the discontinuity is now filled with corrosion products. Initially the zinc corrodes giving Zn^{2+}(aq.) ions, and OH^- ions are formed by cathodic reduction of oxygen at the surface of the cathodically protected steel. These ions then combine to form insoluble $Zn(OH)_2$, although it should be noted that in a natural environment basic zinc carbonates and sulphates rather than $Zn(OH)_2$ are formed. In atmospheric environments the periodic drying of the surface results in the formation of a non-conducting plug of corrosion products, which protect both the surface of the steel and the zinc exposed within the pore. Even if the interior of the pore becomes wet (by condensed moisture or by rainwater) there will be a very high electrical resistance between the steel and the surface of the zinc, so that the galvanic interaction between the two metals will be small.

In natural hard waters a further factor is the precipitation of insoluble calcium carbonate within the pore, resulting from the increase in pH at the surface of the steel and the presence of soluble calcium bicarbonate in the water. This has the same effect as the precipitated zinc salts. A similar situation arises with sprayed aluminium coatings on steel, which owing to the method of formation of the coating result in 'saucer-shaped' particles of the sprayed coating with numerous scattered small pores. Since these particles are coated with a film of aluminium oxide the galvanic action of the aluminium does not become evident until the film has been penetrated. It is considered that initially anodic sites on the aluminium develop at recesses in the pores close to the steel, but the galvanic action between the steel and aluminium does not persist for long since the pores soon become blocked with $Al(OH)_3$ and rust.

It follows from the above considerations that the protective action of anodic coating involves two mechanisms: (a) the sacrificial protection of the substrate, and (b) the formation of insoluble and non-conducting deposits of corrosion products that plug the discontinuities and isolate the substrate from the coating and from the environment. This plugging action may be supplemented by insoluble salts precipitated from the water or by corrosion products of the substrate metal itself.

Cathodic coatings

Although from first principles it would be anticipated that these coatings would stimulate attack on the substrate (*Figure 1.17* (*d*) and (*e*)), which could lead to blistering and exfoliation of the coating (*Figure 1.17* (*f*)), there are again a number of mitigating factors such as those that prevail in the case of anodic coatings. Again the nature of the environment is all important, and whereas there is considerable risk of intensified attack in immersed conditions this is rarely the case in atmospheric conditions. For example, steel covered with a discontinuous coating of nickel will show rusting at pores, but the attack on the steel will be less than if the coating were absent. With cathodic coatings, corrosion at a discontinuity is controlled, as with anodic coatings, by factors such as the conditions of exposure, corrosion potentials of the coating metal and substrate, nature and position of formation of corrosion products (*Figure 1.17* (*h*)), resistance effects, etc. Early work by U.R.Evans showed that a discontinuous coating of copper on steel was not as detrimental as expected, and experiments in which droplets of salt solution were placed on the surface indicated that the resistance of the solution within the narrow pore had a controlling effect. Exposure of these coatings to hydrochloric acid vapour resulted in intense attack on the steel, since in this acid environment the conductivity of the solution is high and the formation of insoluble corrosion products is not possible. On the other hand little attack occurred during atmospheric exposure when the environment was condensed moisture of high resistivity.

Although noble metals such as silver, gold and rhodium are used extensively as coatings for decorative purposes and for certain specialised industrial application, they are seldom applied directly to electronegative metals such as steel and zinc-base diecastings. These metals are normally applied by electroplating, and since they are costly the thickness must be kept to a minimum for most applications; silver plate for cutlery and hollow-ware is an exception to this rule. Gold coatings are used to give a tarnish-resistant finish to silver contacts, but in order to minimise costs they are extremely thin and highly porous. This can result in the formation of corrosion products of the substrate metal, which spread over the surface of the gold and reduce its low contact resistance; silver sulphide formed on a silver substrate is particularly detrimental in this respect.

Nickel and chromium electrodeposits

Nickel and chromium coatings are the most important decorative finishes for a variety of metals including steel and zinc-base diecastings. Both metals are electropositive to steel and chromium is electropositive to nickel, so the galvanic relationship between the coating metals and the substrate at pores is highly complex.

Nickel when plated from a plain Watts bath gives a matt deposit that must be polished mechanically to a bright surface, and although thin deposits may be porous the polishing reduces surface porosity so that attack on the steel substrate is minimal. However, nickel when exposed to the atmosphere forms a dull brown-grey patina; although this is protective to the nickel it is aesthetically unacceptable, so the surface has to be polished periodically. With the advent of chromium plating this difficulty was overcome, and it became accepted practice to apply a thin tarnish-resistant electrodeposit of chromium to the nickel plate; there are still a limited number of applications of nickel plate, but these are usually confined to indoor atmospheric environments or to engineering applications where appearance is a secondary consideration.

Chromium when electrodeposited as a very thin layer is, of course, porous, and at the thickness used in practice the magnitude of the internal stress results in the formation of cracks in the deposit. The surface of electrodeposited chromium when examined under a microscope is characterised by a network of cracks (similar cracks also characterise rhodium electrodeposits).

The next step was the development of bright-nickel plating baths, which produced fully bright or semi-bright deposits requiring no polishing or minimal polishing, respectively. This was achieved by adding levelling and brightening agents to the Watts bath; the resulting nickel deposits contained sulphur, which made the nickel more electrochemically active (lower activation overpotential for dissolution and more negative corrosion potential) than the dull nickel deposited from the plain Watts bath.

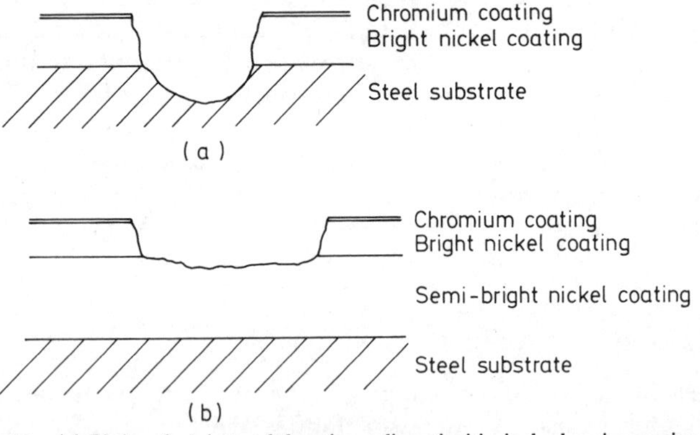

Figure 1.18 (a) Pitting of a bright nickel coating at discontinuities in the chromium coating, which penetrates the nickel and leads to attack on the steel substrate; (b) duplex nickel coating in which the semi-bright nickel coating (low sulphur content) is sacrificially protected by the overlying bright-nickel coating (high sulphur content) so that the former is not penetrated

Chromium when passive is electropositive to nickel. It follows that it will enhance attack on the nickel substrate and that this effect will be greater with the more electrochemically active bright nickel (*Figure 1.18*(*a*)) than with dull nickel. Over the past decade or two a number of improvements have been made to both nickel and chromium plating, and the conventional single layers of nickel and chromium have been replaced by multilayer coatings. The major development as far as nickel is concerned is the use of a duplex system consisting of an initial layer of semi-bright sulphur-free levelling nickel on to which is deposited a sulphur-containing bright nickel in the ratio of 70–80 per cent semi-bright/20–30 per cent bright; this is followed by a conventional or modified decorative chromium deposit (*Figure 1.18*(*b*)).

The chromium produces pitting of the underlying bright nickel; this will proceed until it is penetrated, with subsequent attack on the steel substrate (*Figure 1.18*(*a*)). However, since the bright nickel is anodic to the semi-bright nickel it sacrifically protects it, and attack thus proceeds laterally over the surface; under these circumstances the semi-bright nickel is not penetrated (*Figure 1.18*(*b*)). This results in a characteristic flat-bottomed pit — which is not as unsightly as that produced when the nickel is penetrated and the substrate corrodes, resulting in blistering of the deposit and discoloration of the surface with rust (or with white corrosion products if the substrate is a zinc-base alloy). In sulphur-containing industrial atmospheres the semi-bright nickel becomes activated and behaves in the same manner as bright nickel, so that perforation to the substrate occurs resulting in pits and blisters and exfoliation of the coating; this is particularly prone to occur with zinc-base alloys.

Reasons for applying metal coatings

Metal coatings are applied for two basic reasons: (a) for decorative purposes, and (b) to protect the substrate metal. These two categories are in no way mutually exclusive. A coating that is applied solely in order to protect the substrate metal against corrosion may not in itself be decorative, but one that is applied for decorative purposes will not fulfil its role for any appreciable period unless it provides adequate protection against corrosion of the substrate metal. Thus whichever of the two categories listed above provides the reason for applying a metal coating the true purpose of its use is the control of corrosion.

In any practical application the choice of the material of construction is governed by a number of general factors such as strength, weight, workability and price. Special properties may also be needed such as thermal or electrical conductivity, hardness or wear resistance. When consideration is given to these variables it is frequently found that the most suitable material is one that does not provide adequate resistance to the corrosive effects of the environment. In order to redress the balance and achieve adequate corrosion resistance in service, a metal coating may be applied. Three different examples illustrate this point:

(a) Mild steel is chosen as the material of construction for motor car bumper bars because of its cheapness, its strength and the ease with which it may be mechanically formed to the desired shape. However, its resistance to atmospheric corrosion is very poor; unsightly rusting will develop rapidly and mechanical failure will eventually occur. The application of electrodeposited metal coatings of nickel + chromium to the steel bumper bar produces a finished article with a highly decorative appearance that will be retained for long periods of service, since the coating system will resist atmospheric corrosion with only limited deterioration in appearance.

(b) Steel girder sections chosen for engineering structures such as cranes or bridges may be replaced by aluminium to reduce the weight of the structure. In order to achieve sufficient mechanical strength it is necessary to choose aluminium alloys such as the aluminium–copper–magnesium or aluminium–zinc–magnesium alloy systems. Although pure aluminium offers a high degree of resistance to atmospheric corrosion the presence of alloying elements markedly reduces the corrosion resistance of the resulting alloy. The aluminium–copper–magnesium alloys are susceptible to exfoliation corrosion: this is a special type of intercrystalline corrosion in which attack is preferential along directionally oriented grain boundaries in the metal microstructure and which results in the grains being forced apart by corrosion products, so that they resemble the leaves of a partially opened book (an example of exfoliation of an aluminium alloy is shown in *Figure 1.19*). Many of the

Figure 1.19 Exfoliation corrosion of aluminium alloy (× ½)

aluminium alloys, in particular the aluminium–zinc–magnesium alloys, are also susceptible in varying degrees to stress corrosion (an example of stress corrosion is shown in *Figure 1.20*). If these less corrosion-resistant alloys are given a sprayed metal coating of pure aluminium the good corrosion resistance of the pure aluminium is obtained and the alloy structural member is completely protected against exfoliation or stress corrosion for long periods of service.

Figure 1.20 Typical stress corrosion crack (× 125)

(c) Copper or one of its alloys is likely to be chosen for electrical switch or relay components because of its superior electrical conductivity, but atmospheric tarnishing will lead to an unacceptable increase in contact resistance and cause malfunctioning of the equipment. In order to prevent this, a very thin electrodeposited coating of gold can be applied to the copper switch contacts to exclude the atmosphere and so prevent tarnishing.

The first function of a metal coating, therefore, is the exclusion of the corrosive environment from the substrate metal that it is desired to protect. In doing this it is necessary to consider carefully the reaction between the coating metal and the corrosive environment. It might be thought that the ideal coating metal would offer a greater resistance to corrosion than the substrate material; while this is essential it may not be compatible with the economics of the application (e.g. platinum is an ideal corrosion-resistant coating metal but is impossibly expensive for practical use).

In addition, it is necessary to consider not only the reaction between the environment and the coating metal alone but also the reactions that occur when a metallic couple composed of coating and substrate metals becomes exposed to the environment. This situation can readily arise when discontinuities in the coating expose the substrate through porosity, coating defects, mechanical damage or merely as the result of corrosion of the coating metal itself. If the coating metal is cathodic to the substrate metal when coupled in the particular corrosive environment, the exposure of the substrate at small coating discontinuities establishes a small anode/large cathode relationship, leading to rapid attack concentrated on the small area (as discussed earlier in this chapter). In addition, as corrosion continues the anode/cathode area relationship does not change significantly because the coating is not being corroded and the exposed substrate area also does not increase. Under these conditions the anodic current density remains almost constant, apart from the polarising effects of any corrosion products produced *in situ*, and so corrosion continues at the rapid initial rate.

It may therefore be more practical to use a coating metal that is anodic to the substrate so that sacrificial corrosion of the coating metal occurs, the substrate being protected at any exposed areas with little effect on the overall rate of coating corrosion since the relationship here is one of large anode/small cathode. An example of this type of system is the protection of steel by zinc coatings; the latter are anodic to steel in the atmosphere and completely prevent rusting of the steel until quite large areas are exposed. As sacrificial consumption of an anodic coating continues at a discontinuity the exposed substrate area gradually increases and the cathodic current density, which is already low, decreases still further. In time, this decrease causes the current density to become insufficient to maintain protection at the centre of the increased area of exposed substrate, which will itself then begin to corrode in this region. Sacrificial protection continues to be maintained, however, over the outer regions of the exposed substrate area, which are closer to the large anodic areas of coating remaining.

The rate of sacrificial consumption of an anodic coating may be reduced, and the coating's life therefore extended, if the resistance of the corrosive electrolyte increases, if the substrate exposed at a discontinuity forms a protective film, or if insoluble corrosion products are produced that block the discontinuity. In the case of cathodic coatings, increased electrolyte resistance and the presence of insoluble corrosion products will reduce the rate of attack on the substrate, thus delaying the rate of penetration at small localised corrosion sites. Thus coatings that are either anodic or cathodic to the substrate may be successfully used in practice according to the relative importance of the variables in the particular corrosive environment encountered.

If a coating is to serve a primarily decorative role, its colour, texture and brightness are of maximum importance and must remain stable over long periods of service. This necessitates a high degree of resistance to the corrosive environment such as is achieved, for example, with chromium and with gold. Frequently, in order to combine this stability with adequate protection of the substrate metal, it is necessary to employ a multi-coating system, in which undercoats of other metals are used between the thin decorative topcoat (which may well contain discontinuities) and the substrate metal. The intermediate layers can be chosen to provide adequate protection of the substrate against the action of the corrosive environment, which might otherwise attack it at the discontinuities in the decorative topcoat.

Economic considerations in the use of coatings

The economics of corrosion control are extremely difficult to assess accurately since many factors must be taken into account and their relative importance may vary widely over the lifetime of the article in question. Primarily the economic viability of corrosion control is governed by maximising cost avoidance and revenue gains.

Uhlig[1] produced a general formula for assessing whether a corrosion control process (such as the use of a metal coating) is economically sound:

$$\text{If } 100\,\frac{\Delta T}{T}\left(1 + \frac{L}{C}\right) - 100\,\frac{\Delta C}{C} > 0 \text{ money will be saved}$$

where T = life of structure or component (years)
L = labour costs of replacement of structure or component
C = cost of materials composing structure or component
ΔC = increased cost of using corrosion control process
ΔT = increased life achieved

The annual gain may be calculated by evaluating the above expression and multiplying the result by $C/[100(T + \Delta T)]$.

It will readily be seen that the advantages of using a corrosion control process do not depend on relative processing costs alone, but rather on the relationship between these extra costs and both the improved life expectancy and the costs of replacements. Thus even the use of a relatively expensive process of corrosion control (such as applying a thin coating of an expensive metal to a cheap metal substrate) becomes economically viable if a greatly increased service life is achieved, particularly when the cost of replacement of the structure or component is also high. Three important factors are, however, omitted from the calculation of economic advantage by the above formula.

(a) *Serviceability.* As well as the cost of replacing a defective component, account must be taken of the costs of 'down-time' in any plant or equipment which will occur as a result of failure. Where a component's function is vital to continued operation (such as, for example, a relay switch in a telecommunications or computer complex) the loss of output can be so costly that use of a corrosion control process that is entirely uneconomic by the calculation alone may yet be completely justified. Similar considerations apply also, of course, to those applications where failure cannot be tolerated for safety reasons.

(b) *Application life.* Even if the service-life improvement can be economically justified by the calculation from the formula, the use of a particular method of corrosion control may not be justifiable if it extends the service life beyond the needs of the application. Using the same example as in (a) above, it would be pointless to increase the life of a switch component from, say, five years to ten years if the equipment controlled by it would be in operation for a maximum of only five years.

(c) *Interest charges.* The calculation of cost effectiveness does not take into account the interest charges incurred in providing capital for an operation. It is virtually impossible to quantify this factor in a generally applicable formula because of the ways in which interest rates fluctuate in both the short and the long term. In times of 'dear money' additional expenditure is less justifiable than when money is 'cheap', but account must also be taken of the likely level of interest charges at the time when any replacement will be needed.

All the above three points should be taken into account when calculating the economics of corrosion control for a particular application, and due allowance must also be made for possible fluctuations in labour and material costs over the period under review.

2

Pretreatments

Before any process of metallic coating is applied to a metal substrate it is essential that the latter's surface shall be in a suitable condition to receive the coating. In order to achieve this one or more pretreatment processes must be employed. Broadly, pretreatments fulfil one or more of three purposes:

(a) removal of surface contaminants
(b) removal of superficial corrosion
(c) control of the physical nature of the metal surface

The choice of any or all of these types of process and of the order in which they are applied depends upon the condition of the substrate material as received, on the type of coating process that is subsequently to be used and on the end-use of the coated article.

Because of these differing pretreatment procedures and the factors affecting the choice of specific ones for a given purpose the 'pretreatment line' preceding a coating process can range from a single simple operation to a complex multi-process sequence. It is not the purpose of this chapter to set out in detail the exact requirements of pretreatment lines for particular products and coating processes, but rather to give the reader an outline of the various methods used, the reasons for their use and the ways in which they are applied to meet the requirements of the three classifications given above.

Removing surface contaminants

Surface contaminants are almost always present on materials as a result of production processes that have been carried out prior to receipt, or as a result of deliberate application in order to provide temporary protection or identification. They are, usually, primarily of an organic nature — oils, greases, waxes, paints, lacquers, etc. — but may also be combined with inorganic materials such as metallic debris from the bulk metal produced during mechanical working operations (e.g. swarf or metal soaps) and particulate dirt derived from airborne pollutants.

The presence of surface contaminants always seriously hinders the successful application of a coating process for the following reasons.

(a) They can scar the metal surface during any polishing treatments that may be required, or may even be driven into the surface of the metal so that they cannot easily be removed.

(b) They can provide a physical barrier that will prevent access of a processing solution to the metal surface so that the requisite reaction cannot occur.

(c) They can react with a processing solution, altering its chemical composition and hence its reactions with the metal to be coated.

(d) In the presence of an electrolyte (such as a processing solution) they can react with the substrate metal or with the coating metal, causing corrosion of the surface or producing insoluble products that will further contaminate the surface.

(e) They can be incorporated in a coating system, producing a region where coating adhesion may be defective or interfering with the homogeneity or growth of the coating itself so that a physical defect may develop. Any such defects produced during the early stages of the processing sequence may provide a region where subsequent processing solutions can become entrapped; these pockets of entrapped solution can themselves produce corrosion at a later time.

It follows from the above possibilities that surface contaminants should always be removed before subsequent processing is attempted. The principal way in which contaminants of this nature are removed from metals is by the use of cleaner-degreasers.

In its simplest form a cleaner-degreaser may be merely a tank of organic solvent (such as carbon tetrachloride, benzene, toluene, acetone, etc.) maintained at room temperature, into which the work may be dipped or swabbed. Oils, greases and lacquers are softened by the solvent and taken into solution, and entrapped insoluble dirt and metal particles are loosened so that they can fall away to the bottom of the vat. However, simple immersion or swabbing in the cold is an inefficient means of cleaning all but limited quantities of small articles. Problems are associated with the extraction of toxic vapours from the solvent; also the vat quickly becomes contaminated with dirt and greases removed from the work, which form an emulsion that is retained as a film on the metal surface after removal and drying.

Heat can be applied to the bath to accelerate cleaning action, and some amelioration of bath contamination can be achieved by working a closed-circuit flow of solvent incorporating settling and/or filtering, but even these methods remain of limited efficiency.

The most commonly used method of solvent degreasing that operates at high efficiency is the hot liquid/vapour degreaser plant. The principle of this type of equipment is shown in *Figure 2.1.* The solvent is contained in a tank in which it can be heated to boiling point. There is an annular compartment in the upper part of the side walls of the tank within which is a cooling coil where the hot solvent vapour is condensed back into liquid. This condensate is collected in the bottom of the annular compartment, from which it flows back by gravity feed into the main liquid compartment below to recommence the vaporising-condensing cycle. The tank is covered and vented through a vertical flue, which extracts any uncondensed fumes that may escape from the upper portion of the tank.

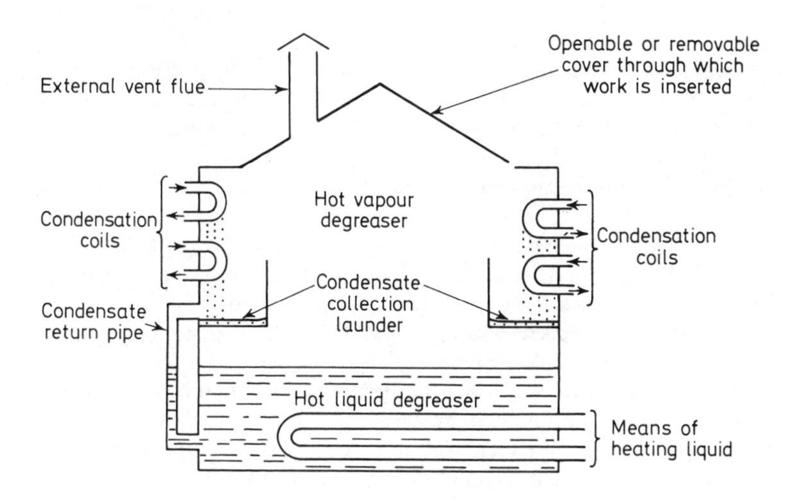

Figure 2.1 Diagrammatic sketch of hot liquid/vapour degreaser

If cold, soiled work is introduced into the upper part of the chamber where the solvent vapour is present at a temperature of, say, 87°C (the boiling point of trichloroethylene), condensation of the solvent takes place on its cool surface. This continues until the temperature of the work attains that of the solvent vapour. During this period the continuously renewed flow of condensate on the surface of the work flushes away soils and greases, which fall to the bottom of the tank. If the work is heavily soiled with resistant contaminants, treatment in this way by solvent condensate may be insufficient to effect complete cleaning. In such cases the work may be totally immersed in the boiling solvent tank in which the heavy soils will be removed with a high degree of efficiency. After total immersion a light grease film may be retained on draining, cooling and drying, but this can readily be removed by a subsequent treatment in the vapour compartment of the degreaser. Provision is also made in the equipment for draining the liquid compartment from time to time in order to remove any build-up of soil sediments and so retain the high efficiency of the equipment.

The solvents used in this type of plant are the chlorinated hydrocarbons, the most commonly used being trichloroethylene. This solvent does not itself attack most metals but a violent reaction can occur between the hot solvent and finely divided light alloy metals, so precautions must be taken when handling these materials. Cases have been known where highly destructive explosions have resulted from the introduction into vapour degreasers of light alloy components carrying fine swarf or metal dusts. Special formulations of solvents containing additives to increase their stability have been developed, however, so these alloys can be safely treated. Care must also be taken to avoid a build-up of acidity in the solvent, since an acidic solvent may readily attack the metal articles being cleaned and, in serious cases of acidic build-up, the materials of construction of the plant itself. Wet work cannot be treated in a trichloroethylene plant, but if perchloroethylene is used as an alternative solvent wet work can be safely treated. In addition, perchloroethylene has a higher boiling point (121°C) than trichloroethylene and this leads to improved efficiency of removal of hard greases, although at the expense of greater heating costs.

As an alternative to (or in a number of cases in combination with) solvent cleaning, soils and greases can be removed by chemical cleaning methods. Chemical cleaners can act on the soils in a number of different ways such as solubilising, emulsifying, saponifying and peptising. An alkali detergent-powder mixture is the most commonly used basis for chemical cleaners.

Alkali metal silicates, phosphates and carbonates are employed as hot aqueous solutions; the addition of surface active agents serves to lower surface tension so that the soiled work is more readily wetted by the cleaner, and promotes emulsification of oils and greases. The alkali salts themselves have good detergent properties, causing saponification by reaction with fatty substances and promoting peptisation which assists the retention of insoluble soils in suspension in the cleaner. Sodium metasilicate and trisodium phosphate are among the most commonly used of the alkali salts, but they are often fortified in purpose-formulated cleaners by additions of tripoly- or hexameta-phosphates, which chelate hardness salts in the make-up water and prevent precipitation of insoluble deposits on the work or plant.

Although caustic solutions are more efficient saponifiers than silicates, phosphates and carbonates they will react with many metals — notably the light metals and alloys — and they are considerably more difficult to rinse from the surface of the work after treatment. Some free caustic may, however, be incorporated in heavy-duty cleaners.

Finally, buffering agents may be added to the formulation of alkali cleaners.

Treatment by alkali cleaners may be effected by immersion in soak tanks, efficiency being improved by agitating either the liquid or the work, or by spray application from pressure jets. Thorough water rinsing must always be subsequently employed.

The efficiency of chemical cleaners can be much increased, and the danger of chemical attack on the metal reduced or prevented, by electrolytic action. A polarising current of $\sim 500\ A/m^2$ at an applied voltage of 3–12 V is used, the work being made either anodic or cathodic according to the metal concerned. Ferrous metals are anodically cleaned and copper-base materials are treated cathodically; in many cases a brief reversal of polarity is employed prior to removal of the work from the cleaner so as to remove any electrodeposited smuts. The cleaning action of the process depends on the formation of gas bubbles on the surface of the work as a result of the discharge of hydrogen or oxygen gas at the metal surface. The bubbles of gas are formed at the metal surface beneath the soils and provide a mechanical removal action. In addition, cathodically produced alkali improves detergent action. Electrocleaning is not suitable for the treatment of tin, lead, zinc, aluminium or light alloys.

Both solvent and chemical cleaning may be assisted and their efficiency improved by employing ultrasonic agitation of the work while it is immersed in the cleaning liquid. A transducer built into the liquid tank induces ultrasonic vibrations in the immersed work and bubbles of gases or cavitation bubbles are produced at the work surface. When these bubbles either form or collapse, mechanical loosening or removal of the soils attached to the surface takes place thus improving the efficiency of the cleaning process.

Removing superficial corrosion

Superficial corrosion of metals occurs as a result of oxidation during processing (e.g. in hot-working processes and in heat-treatment processes) or through reaction with a corrosive environment during storage. Although the extent of this corrosion can be controlled and minimised by appropriate control during processing and by the use of temporary protective measures during storage it is extremely unlikely that it can be completely prevented. Any corrosion products on the metal surface must be completely removed before coatings are applied since their presence interferes with the application and/or the performance of the coating. Loose or brittle oxide films entrapped between the coating and the substrate produce regions of poor adhesion where breakdown of the coating can easily occur in service. Corroded areas may not be receptive to electrodeposition so that bare areas remain after plating, and the difference between the electrode potential of a corroded region and that of the rest of the substrate can produce local electrochemical action leading to enhanced corrosion in service.

The removal of unwanted processing or storage corrosion may, of course, be effected by mechanical means during machining, polishing or abrading, which are discussed in the next part of this chapter. Apart from the use of these methods removal of corrosion is generally achieved by chemical immersion treatments known as pickling.

The pickling process is in many ways akin to the chemical cleaning processes already described — indeed, pickling is only a more aggressive form of chemical cleaning aimed not at greases or soils but at oxides and other more stable metal compounds. Removal may be achieved either by solution of the corrosion products in the pickle liquor or by their physical detachment from the metal surface when they are undermined by chemical attack on the substrate.

The aggressivity required of a pickle makes it necessary to move from the nearly neutral or mildly alkaline salts used as chemical cleaners to stronger acids or alkalis. The concentration and operating temperature are increased as the pickling duty moves from the removal of light tarnish stain to the removal of heavy oxidation and scaling. Once again specific formulations may involve the use of wetting agents to improve efficiency and speed of action, and inhibitors to reduce, or even completely prevent, attack on the clean metal beneath the corroded surface.

A different approach to pickling that is of particular benefit in the case of heavy, tough and adherent scales is the use of molten salt baths. The removal action in this type of pickling may combine chemical attack on the scale by the molten salt with a shattering of the continuity of the scale by differential expansion from the underlying metal as a result of the thermal shock of immersion in the molten bath. This method of pickling is finding increased application in a number of fields and may be of particular benefit as a way of combining descaling and heat-treatment in a single operation. However, the process requires special equipment and skilled operators, is costly and may be hazardous. In addition, because of the very nature of molten salt pickling the process cannot be employed where exposure to high temperatures will adversely affect the mechanical properties of the metal to be descaled. Molten sodium hydroxide and molten sodium hydride (NaH) are frequently used for this purpose.

As with chemical cleaning, the action of pickling may be assisted by electrolytic action (using either anodic or cathodic polarisation of the work) or by the use of ultrasonic agitation.

The use of different types of pickling treatments for the various metals is summarised in *Table 2.1*.

Table 2.1 SUMMARY OF PICKLING METHODS FOR DIFFERENT METALS*

Metal	*Soak cleaning*	*Immersion pickling*	*Electrolytic pickling*	*Salt-bath descaling*
Iron or steel	Dilute acids used for removing light corrosion only. Pitting can occur with cast iron	Simple acid solutions used for removing rust or scale from plain carbon steels or cast irons. Stronger acid mixtures used for alloy steel. High-strength steels may suffer hydrogen embrittlement. Cast irons may become pitted	Anodic or cathodic treatment in acids used for steels especially prior to electroplating. Alkaline processes suitable for treating cast iron.	Mainly used for removing heavy scales from alloy steels and for removing siliceous scales from cast iron
Copper-base alloys	Dilute sulphuric acid used for removing light tarnish	Dilute mineral acids, often in mixtures or with addition of dichromate salts, used for removing heavier oxide scales	Mild cathodic alkali processes used for removal of light tarnish	Mainly used to remove very tough scales or adherent siliceous scales
Zinc and its alloys	Very dilute acids only used with short duration treatments		Not used	Not used
Tin and lead	Dilute acids used for removing light tarnish	Fluoboric acid solutions used for general pickling	Not used	Not used

Table 2.1 *(continued)*

Metal	*Soak cleaning*	*Immersion pickling*	*Electrolytic pickling*	*Salt-bath descaling*
Aluminium and its alloys	Dilute acid or alkali solutions used for light etching only. Smut deposits removed by subsequent nitric acid dipping	Nitric/hydrofluoric acid mixtures and hot chromic/sulphuric acid mixtures used for general pickling. Hydrofluoric acid or caustic alkali mixtures used for etching	Not used	Sodium hydride used for removing adherent siliceous scales
Magnesium and its alloys	Not often used	Chromic/hydrofluoric, nitric, phosphoric, acetic and sulphuric acids all used in combinations for general pickling and etching	Not used	Not used
Nickel and its alloys	Not used	Sulphuric and hydrofluoric acids used for general pickling	Cathodic treatment in acids	Little used except for heat-resisting high-nickel alloys
Titanium	Not used	Sulphuric acid used for removing light scale. Fluoboric, hydrofluoric and nitric acids and mixtures used to remove heavier scales	Not used	Frequently used for removal of very heavy scale. With caustic salts treatment temperature must not exceed 480°C

*Based on data in *Finishing Handbook and Directory*, Sawell Publications Ltd (1970).

Controlling the physical nature of the surface

Primarily, the required condition of the surface of a metal to be coated is governed by the end-use of the finished product. Most coating processes can be applied equally well to cast, wrought, polished or

roughened surfaces provided always that these surfaces h
thoroughly and scrupulously cleaned as indicated in the
sections.

The one notable exception to this generalisation is the case of coatings applied by metal spraying processes (see Chapter 3). The way in which sprayed metal coatings are built up is such that in order to achieve adequate adhesion between the sprayed coating and the substrate the surface of the latter needs to be roughened so as to provide a mechanical keying action to retain the coating during service. The degree of roughness and the angularity of the surface irregularities both markedly affect the adhesion, and it is also important to ensure that the roughened surface is free from contamination.

Pretreatment for metal spraying, therefore, is accomplished by grit-blasting, taking care to see that the range of grit sizes used is carefully controlled. Too fine a grit produces a surface with insufficient roughening for adequate adhesion; too coarse a grit produces an unacceptable degree of macro-roughening while probably still having insufficient micro-roughening to achieve the optimum coating adhesion. The actual range of grit sizes used depends on the materials of which the grit is composed, upon the metal that is to be treated and also, to a lesser degree, upon the air pressure supplied to the grit-blasting equipment. Chilled iron and alumina grits are the two materials most commonly used. In the interests of economy of materials it is usual to collect the grit and recycle it for further use; this is usually done by means of suction pipes placed adjacent to the treated surface during blasting or by carrying out the blasting operation in an enclosed cabinet from which the grit is collected and piped back to the blast nozzle (*Figure 2.2*). Care must always be taken to remove dirt and excessive fines produced when larger sizes of grit shatter in use.

The freshly abraded metal surfaces produced by grit-blasting tend to be chemically active and thin, air-formed oxide films are readily formed on them. For this reason operators should not handle the grit-blasted surfaces without using gloves and the sprayed metal coatings should be applied as rapidly as possible after grit blasting since any deterioration will adversely affect the performance of the coating. The delay between grit blasting and metal spraying should never be sufficient to allow visible deterioration of the surface to occur; the time limits to avoid this vary with the conditions under which the operations are being carried out. Specifications for metal spraying lay down maximum permissible delays between grit blasting and metal spraying; e.g. Defence Standard 03–3 (Protection of Aluminium Alloys by Sprayed Metal Coatings) allows a maximum of four hours under good workshop conditions and suggests that in on-site applications the delay should not exceed a few minutes.

Apart from its use as a pretreatment for metal spraying, the grit-blasting process may be used for materials subsequently coated by other methods. In these cases it is used to remove heavy scale from a metal prior to the employment of other pretreatment processes or to provide a surface with a controlled degree of roughness that may be required for decorative or frictional purposes. For achieving the finer grades of roughening — known as satin finishes — the process of

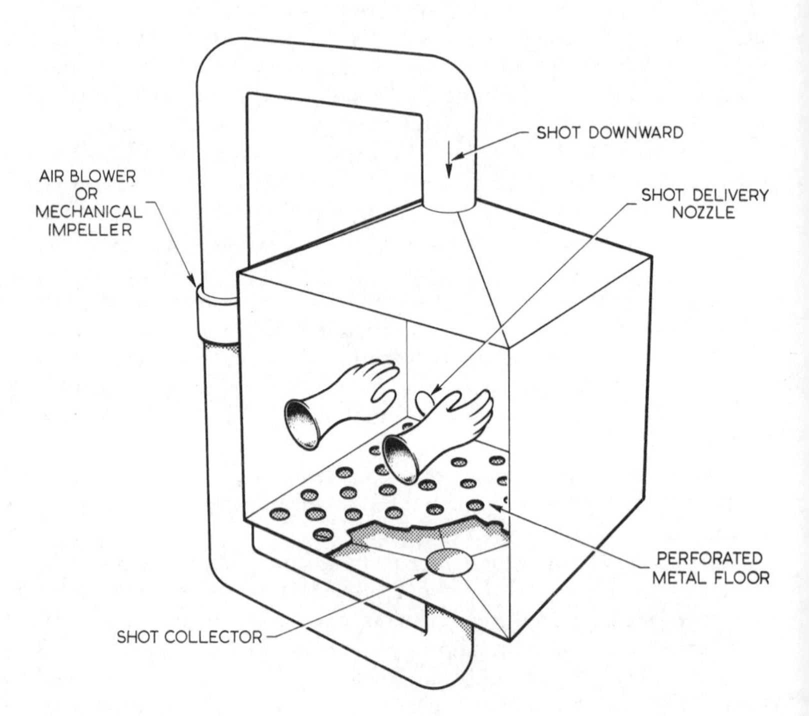

Figure 2.2 Grit blasting cabinet

vapour blasting may be used; this is essentially the same type of process as grit blasting except that very fine abrasives are used and are applied to the work by means of a pressure jet of water vapour.

Where processes other than metal spraying are used to apply metal coatings to cast, wrought (i.e. rolled, forged or as extruded) or machined surfaces it is only necessary to ensure that greases and soils or oxide films and scales are removed by using the appropriate cleaning techniques previously discussed. Three other classes of surface finishes may, however, be required — abraded, polished and etched. These conditions are achieved in the following ways.

Abrading or grinding

Abrading consists of the controlled removal of metal from a surface by the application of grits of graded coarseness or by the use of rotary wire brushes. The grits may be cemented to paper, cloth or metal bands, strips or discs, and usually consist of tungsten carbide, alumina, diamond or siliceous materials supplied in a range of carefully controlled coarsenesses. The abrading process may be carried out by hand or by mechanical equipment and performed either in the dry state or lubricated with water or oils. Some degree of macro-levelling of the surface is effected, but a micro-roughened finish is produced that may be directionally or randomly oriented according to the way in which the process is applied. The pressure used to apply the abrasive to the work and the type and extent of lubrication used must be carefully controlled to avoid embedding particles of metal debris into the surface, where their presence could lead to the formation of defects in subsequently applied metal coatings.

Abrading or grinding may be used prior to polishing or as finishes in their own right. In the latter case they must always be followed by degreasing and/or cleaning treatments in order to remove soils and metal dusts before applying metal coating processes.

Polishing

Polishing is used to improve appearance, levelling, reflective properties or closeness of fit with mating components or to reduce friction between moving components. Mechanical, chemical and electrolytic methods may be used.

Mechanical polishing

This method may be considered as an extension of the abrading process at the finest end of the scale; metal removal is reduced and smoothing action accentuated. When applied in its final stages to produce the greatest lustre and smoothness it is known as buffing.

Very fine grades of emery, alumina or silicon carbide are used as polishing abrasives and are applied to the work by means of resilient felt or cotton mops or wheels. Tripoli and rouge may be used for obtaining the finest lustres. The abrasive may be fixed to the wheel with glue or retained in position by tallow or grease compounds used to lubricate the polishing process. Lubrication serves to assist metal flow (and hence smoothing) and to prevent gouging and embedding of abrasive particles in the metal surface. Localised heating of the surface during polishing, caused by friction, can also assist the polishing action.

Great care must be exercised when polishing the softer metals, in which excessive metal flow and smearing can occur. A particular example is zinc diecastings, where sub-surface porosity can occur beneath the casting skin; if excessive metal flow is allowed to occur the sound cast skin may be broken to reveal the porosity holes, with consequent bad appearance and the danger of entrapment of subsequent processing solutions.

After mechanical polishing very thorough degreasing and/or chemical cleaning must be carried out to ensure complete removal of particles of metal and abrasive and the removal of greases or waxes used for lubrication.

Mechanical polishing and burnishing can also be carried out by treating the work in rotating barrels or vibrating tubs. The articles to be polished are loaded in the containers together with ceramic or metal shapes or chips, polishing compounds, and water to act as a lubricant. Chemical buffering salts and wetting agents may also be added. The rubbing action between the work and the chips during rotation or vibration of the containers enables the polishing compound to act to remove metal from the surface of the work and so produce smoothing and brightening. Careful control of the components of the mix in the containers and the total loading and rate of rotation or vibration enable an optimum of polishing to be achieved without mechanical damage to the work or loss of detail in shaped parts.

Chemical polishing

The amount of true polishing that can be achieved by chemical immersion treatment is limited, and the process would perhaps be more accurately described as chemical brightening. Thus it is not possible to achieve a true mirror surface, though some degree of smoothing does occur and general brightness and reflectivity are improved.

The essence of the process consists of acid dissolution of the metal from the surface, the rate of attack being limited by controlling the rate of diffusion of soluble salts from the surface and the replenishment of free acid in the region. This is normally achieved by increasing the viscosity of the polishing solution and adjusting the formulation so that it contains large, complex molecules. Under slow rates of diffusion replenishment of fresh acid is slowest in deep recesses on the surface of the work and most rapid on asperities. Consequently, more metal is removed from the high spots of the article and a degree of micro-levelling may be achieved.

Most of the commercially available chemical polishing solutions rely on a chemical such as ortho-phosphoric acid to increase viscosity, the active reagent for metal dissolution being an oxidising acid such as nitric acid. Buffering agents and other salts to control dissolution rates may be included in the formulation. The process is usually operated at an elevated temperature, and work may be treated either individually or in batches contained in baskets constructed from materials resistant to the action of the polishing solution. Considerable quantities of toxic fumes are produced and must be efficiently extracted. The work must be rinsed very quickly and thoroughly after treatment since any polishing solution retained on the surface of the work after removal from the bath will continue to react with the metal until complete exhaustion occurs, thus leading to uneven results.

The only pretreatment required for chemical polishing is the complete removal of greases or other adherent soils to ensure that the surface of the metal is completely wetted by the polishing solution. However, the limited extent of the polishing action achieved makes it essential that a fairly high-quality surface finish be obtained before applying a chemical polishing process.

Some limited degree of regeneration of chemical polishing solutions can be achieved by the addition of carefully controlled quantities of the active reagents consumed in the process, but the build-up of metal salts in the solution has a maximum tolerance level after which polishing actions becomes progressively reduced. A further problem with chemical polishing is the active state of the metal surface after processing; this is revealed by the greater rapidity with which light tarnish develops, and in order to avoid deterioration from this cause it is necessary to apply any subsequent coating process immediately or to make use of temporary protectives.

Aluminium and copper and their alloys are the commonest metals treated by chemical polishing techniques; formulations are also available for the chemical polishing of silver.

Electropolishing

A characteristic of solutions used in chemical polishing is their high redox potential, which is due to the addition of powerful oxidising agents such as nitric acid; during polishing these are cathodically reduced at a high rate with concomitant rapid anodic dissolution of the metal. On the other hand, in electropolishing the metal is made the anode of the cell in which the cathodic reaction occurs at another electrode — the cathode (an inert conductor such as platinum, stainless steel, carbon, etc.). Thus, whereas in chemical polishing the potential is controlled by the redox potential of the solution (limited to ~ 1.1 V versus SHE), much higher and more controllable potentials can be achieved in electropolishing by the e.m.f. supplied by an external source. The solutions used are far less aggressive than those used in chemical polishing and are frequently reducing acids, although again they are formulated so that they are viscous.

Although many of the theories to explain electropolishing invoke the concept of diffusion-controlled dissolution, Hoar (*Nature,* **165,** 64, 1950) was one of the first to propose that crystallographic etching (without polishing) was suppressed by the formation of a thin compact *solid* film on the surface. Under these circumstances the anodic process is determined by the random appearance in the solid film at the metal/film interface of cation vacancies into which random metal cations can pass. This random dissolution gives a smooth micropolished surface.

Thus when a metal is made anodic in a suitable electrolyte, dissolution begins to occur. At low current densities dissolution is crystallographic and the surface becomes etched, but as the current density is increased a critical range is entered within which the diffusion layer is produced adjacent to the anode and reaches a maximum thickness.

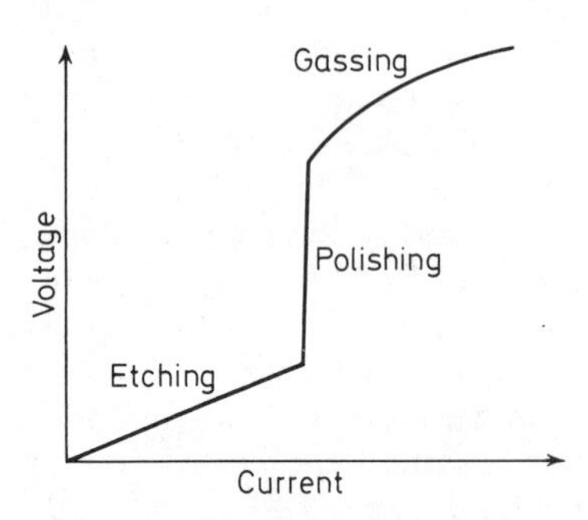

Figure 2.3 Effect of voltage–current relationship on electropolishing

The process then comes under diffusion control and in this range of operating conditions polishing action occurs, with true microlevelling of the surface so that mirror finishes can be obtained. Any further increase in current density leads to gassing at the anode and polishing action is lost. See *Figure 2.3,* and compare with *Figure 1.15.*

A wide range of solutions is available for electropolishing metals, each being specific to a particular metal or group of metals or alloys.

Generally these solutions are much less concentrated and hence more safely handled than those employed for chemical polishing, and toxic fumes are not normally evolved during electropolishing processing. Electropolishing solutions tend to have a longer active life than chemical polishing solutions since in many cases metal taken into solution from the work is plated out on the cathode. Most of the ferrous and non-ferrous metals and alloys can be readily treated by commercially available electropolishing solutions.

As with chemical polishing, a clean and wettable surface is the only prerequisite of treatment by electropolishing techniques. However, whereas chemical polishing can be undertaken on batches of components, all items for electropolishing must be individually jigged or wired so as to provide the necessary electrical connections and this adds considerably to the operating costs of the process.

In chemical polishing, apart from the micro-smoothing, metal removal occurs uniformly over the whole of a shaped surface, whereas in electropolishing metal removal is less in very deep recesses and is minimal on the back of the article, which is shielded from the cathode. This feature of electropolishing can be exploited where it is desirable that metal removal should be limited in certain areas of shaped work.

Rinsing after electropolishing must be thorough but need not be so rapidly carried out as with chemical polishing, since the solutions used for electropolishing generally only attack the metal minimally in the absence of the polarising current. Furthermore, electropolished surfaces tend to be somewhat more tarnish-resistant than chemically polished surfaces, so subsequent processing does not need to be carried out so rapidly.

Etching

Etching of metal surfaces may be chosen as a pretreatment process in order to produce satin-finish decorative effects, or to produce micro-roughening where a key is necessary to improve the adhesion of any subsequently applied coatings. Since the action is one of controlled roughening (or 'anti-levelling') it follows that the necessary degree of macro-smoothness of the contours of the work must be achieved by controlling the manufacturing processes or by using a grinding or part-polishing process before the etching stage is applied.

Etches always work by random dissolution of metal from the surface, as judged on a micro-scale, but their action is frequently highly selective on a micro-scale since individual grains of the microstructure of the metal may be attacked or inert according to their orientation. The action may be wholly chemical, chemical assisted by anodic electrochemical action, or wholly anodically electrochemical.

Both acidic and basic solutions may be employed in chemical etching processes according to the metal to be treated. Aluminium and its alloys are commonly etched in caustic-based solutions, to which may be added buffering and wetting agents, sequestrants, and a range of salts to control the severity of the etch and to complex the aluminium ions. Alternatively, acid-based solutions may sometimes be used. Copper-base alloys and ferrous materials are normally etched in solutions of the oxidising or mineral acids that are also employed for pickling processes, but the concentration of acid is generally less than that used in pickling; metal salts are commonly added to the etches to produce a partial inhibition of metal removal or to modify the dissolution process so that it becomes more selective on certain features of the metallurgical structure of the metal.

Anodic etching of ferrous materials is normally carried out in a sulphuric acid solution, the concentration required being inversely proportional to the severity of the etch required. The anodic current density and the treatment time are increased in order to obtain deeper etching effects, and gassing occurs at the anodes. Copper may be anodically etched at high current densities in a solution of mixed chlorides; the process is widely used for halftone printing plates.

Etching is widely used (particularly in printing, engraving and electronic applications) for the selective removal of metal from closely defined areas and to produce raised or recessed patterns. In order to achieve these effects the metal surface is first coated with a wax, or other material resistant to the etching solution, in those areas where removal of metal is not required; etching is then confined to the uncoated portions of the metal surface. Finally, the wax coating is removed by solvents after etching and rinsing have been completed.

In order to obtain uniform etching, metals to be treated must be free from scale, soils and greases and must be completely wettable by the etching solution. After etching thorough rinsing must be carried out; as with chemical polishing, this rinsing should be done as rapidly as possible in order to prevent etchant retained on the surface from continuing to react with the metal in localised regions.

3

Coating processes

A variety of processes exists for applying metal coatings that are to be used for corrosion control. They cover a wide range of different techniques and produce coatings with differing characteristics according to their method of production, although there are only limited variations in the corrosion behaviour of a given metal applied by different coating processes. The broad classification of coating processes covers the following groups:

(a) molten application
(b) spray application
(c) chemical deposition
(d) electrodeposition
(e) vapour deposition
(f) diffusion application
(g) mechanical application

Within each of these groups there are several ways of producing the required coatings, depending upon the nature and design of the substrate material. The type of coating produced and its properties also depend upon the coating/substrate combination.

Molten application

In molten application the coating metal is heated to the molten state and the substrate material is either immersed in the molten bath (as in hot-dip processes) or has the molten coating metal flowed or otherwise transferred on to its surface (as in soldering processes).

In order to achieve complete coverage of the surface by the coating metal and an adequate bond between the coating and the substrate, it is necessary to ensure that the oxide film is removed so that the molten coating metal completely wets the substrate; this is achieved by fluxing. When the flux becomes molten it dissolves any oxide that may be present on the surface of the substrate material, so producing a completely clean surface to receive the molten coating metal. At the same time any traces of oxides in the molten bath are also fluxed away. Under these conditions complete wetting of the substrate by the coating metal can take place.

The molten coating metal may react with the solid substrate metal to produce an alloy by diffusion, the composition and extent of the alloy layer varying with the constituent metals and with the time of treatment at the elevated temperature. Although, in general, the thickness of the alloy layer increases with time at temperature, the reaction is not linear with time and there are practical limits to the growth of any alloy layer that can be obtained. Furthermore, as thicker alloy layers are produced the alloy composition across the thickness of the layer will vary, becoming progressively richer in the coating metal with increasing distance from the interface.

On removal from the molten bath a top coating of the pure coating metal is retained on the surface of the alloy layer; its thickness is governed by the fluidity, surface tension and rate of solidification and, to a lesser degree, by the rate of removal of the work from the molten bath. Close control of dimensional tolerances is difficult to achieve when using molten coating techniques. There is, of course, a tendency for a build-up of coating thickness to occur in recessed areas and on the lower draining edges; conversely, thinner coatings are found on the peaks of protruding portions of the surface due to the flow of metal away from these regions before solidification is complete. These thickness variations can be minimised by increasing the rapidity of cooling to speed up the solidification process and (in the case of continuously processed strip or wire) by passing the emerging work through rolls or wipers.

Since the process of hot-dipping requires immersion of the work in a molten bath of the coating metal its use is restricted to materials in which the melting point of the substrate metal is considerably higher than that of the coating metal. Account must be taken of the fact that during processing the substrate metal will be annealed. In the case of soldering (where the application of heat for the coating process may be localised to some degree) annealing effects may be minimised, but nevertheless the effects of annealing must always be considered when designing an article to be coated by molten application. Heat distortion can occur with material of thin section or with sections of varying thickness, and also with assemblies, particularly where constructional stresses are present. The heat-distortion effect on castings of varied thickness can, in extreme cases, lead to the occurrence of fractures. It is often more desirable to dip-coat individual components and assemble them after coating rather than to coat the completed assembly.

Articles suitable for hot-dip coating can range in size from components such as small fasteners, which are treated by batch processing in drainable containers, through larger components and rigid sections to assemblies, which are individually suspended in the molten bath. Limitation of the size of components for individual handling is exercised by the maximum dimensions of baths available to contain the molten coating metal and the capacity of the lifting gear employed. Components up to 18 metres in length can be readily treated in a single dipping operation in, for example, hot-dip galvanising plant which is commercially available; longer components — up to, say, 30 metres in length — can be coated by employing double end-dipping techniques. Sheet, strip or wire can be treated in an automated, continuous processing line (often at high speeds) by looping

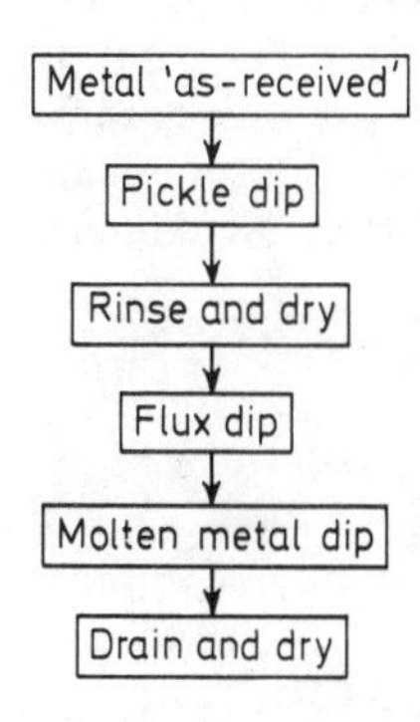

Figure 3.1 Flow-chart of hot-dip metal coating operation

successively through pickling, fluxing and coating baths, after which sizing rolls may be used for dimensional control. Post-coating treatments may also be incorporated in the processing sequence.

When designing for hot-dip coating, deep, sharp corners should be avoided and large fillet radii used. Allowance must be made for build-up in thickness as the coating is produced by providing suitable clearance in the sizing of mating and moving components. Provision must always be made for allowing the molten metal to drain freely from recesses and hollow sections after removal from the bath; thus drainage holes should be provided and blind holes avoided. Holes and hollow sections of small bore must be avoided as they may become filled with solidified metal. Totally enclosed hollow sections must always be adequately vented since expansion of air contained within them, which occurs on immersion in the molten bath, could otherwise lead to explosive rupturing.

The pure coating metals applied by hot-dipping are always softer than the substrate metals to which they are applied. Consequently this method of coating is seldom suitable for applications in which resistance to wear is required. Although the pure coating metals are soft, alloying with the substrate metals increases hardness and decreases ductility; the alloy layers also frequently exhibit corrosion resistances different from those of the pure metals.

The processing sequence for hot-dipping is shown in outline in *Figure 3.1*, but in order to consider the preparation and properties of the various hot-dipped coatings in greater detail it is necessary to consider the different processes individually.

Hot-dipped zinc on steel (galvanising)

Steel for galvanising is first pickled in a solution of hydrochloric acid to remove all rust and scale and to roughen the surface lightly. The pickling acid usually contains organic inhibitors, which prevent excessive attack on the clean steel while allowing reductive dissolution of oxide films and scales. Castings should be given a preliminary gritblasting treatment. After pickling the metal is fluxed with

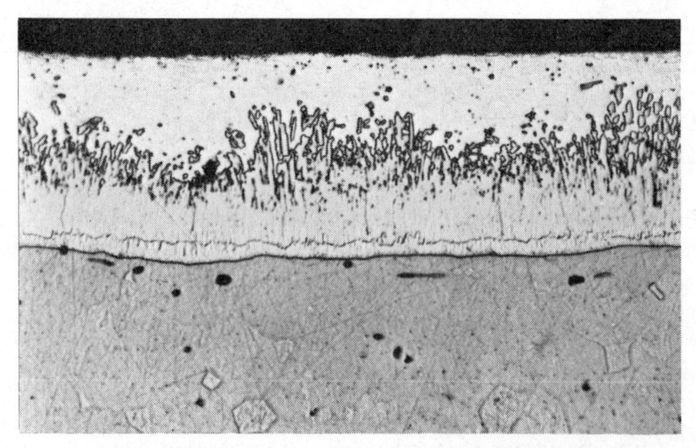

Figure 3.2 Hot-dip galvanised coating on steel (× 400)

ammonium chloride, either as a separate operation before immersion in the molten zinc or by means of a molten flux cover on the zinc bath (or in some cases by using both methods).

During immersion in the zinc bath, which is usually operated at a temperature between 430 and 470°C, zinc–iron alloy layers are formed. Three alloy compositions occur, containing 6.25 per cent (ζ phase), 11 per cent (δ phase) and 22 per cent (γ phase) iron respectively. The alloy richest in iron is formed in the region of the coating adjacent to the steel substrate, that lowest in iron adjacent to the pure zinc outer coating. A typical, good-quality galvanised coating on steel in which the alloy layers can be clearly seen is shown in *Figure 3.2.*

A number of defects can occur in galvanised coatings, any of which can affect their corrosion protection performance. Operation of the galvanising bath at excessive temperature causes a reduction in the thickness of the pure zinc top coating, and a sharp rise in the rate of alloy formation occurs at bath temperatures above 480°C. The production of excessively thick alloy layers decreases the ductility of the coating as a whole, because of the more brittle nature of the zinc–iron alloys, and the coating may delaminate from the substrate if the coated article is subsequently bent. Furthermore, the thinner layer of pure zinc has a reduced capacity to provide sacrificial corrosion protection to the substrate in service. The addition of a small percentage of aluminium (0.1–0.2 per cent) to the molten zinc bath substantially decreases the rate of alloy formation and improves the ductility of the coating.

During operation of the galvanising bath, oxidation of the zinc to zinc oxide increases its viscosity and tends to embrittle the coating. Dross (a zinc–iron alloy containing 3–7 per cent iron) is also formed in the bath; it is pasty at the galvanising temperature and, being denser than the molten zinc, tends to drop down through the bath. If particles of dross settle on the immersed work the normal coating process is interrupted and rough, defective areas are produced on the finished article.

Although the zinc–iron alloy layer formed during galvanising is more brittle than the pure component of the coating, the alloy has a slower corrosion rate than the pure metal. Advantage may be taken of this slower corrosion rate to improve service performance by post-galvanising annealing, during which the thickness of the alloy layer increases by continued solid-state diffusion until, ultimately, the complete coating can be converted into alloy.

Coatings produced by hot-dip galvanising are normally specified by weight per unit area, i.e. the coating on both sides of the substrate is taken into account; typical coating weights are in the range 20–50 g/m^2, corresponding to thicknesses in the range 10–30 μm. In exposure to a corrosive environment zinc is slightly anodic to the zinc–iron alloy layers, and both are markedly anodic to the steel substrate. For these reasons the coating will be attacked preferentially to the substrate and provide sacrificial protection at any areas of the substrate that may be exposed at coating discontinuities (see Chapter 1). Complete prevention of rusting of the steel substrate will be achieved by this means for periods of exposure to the atmosphere ranging from about two years in severely polluted industrial regions to perhaps as much as fifty years in mild rural environments.

A number of reasons exist for the great differences in sacrificial life in different environments, but that which exerts the greatest influence is probably the nature of the zinc corrosion products. In industrial atmospheres polluted with sulphur compounds zinc sulphates are formed; these are soluble and are removed from the surface by rain, allowing corrosion of the zinc to continue freely. In rural and marine atmospheres, however, basic carbonates and chlorides are produced as corrosion products. These salts are less soluble than the sulphates, and partial stifling of the corrosion of the zinc occurs. The production of less soluble basic carbonates and chlorides is also responsible for the limited rate of consumption of the sacrificial zinc coating, which enables galvanised steel to achieve a very long rust-free life in applications involving immersion in many natural waters.

Hot-dipped aluminium on steel (aluminising)

This process closely resembles that of hot-dip galvanising in that pickled and pre-fluxed steel is immersed in a bath of molten aluminium, with which it reacts to form aluminium–iron alloy layers and a pure aluminium topcoat retained on removal from the bath. However, the process is more difficult to operate and control than that of galvanising owing to two major factors. These are the higher melting point of aluminium and its greater rate of oxide formation compared with zinc. In order to achieve sufficient fluidity in the molten aluminium the operating temperature must be maintained above 700°C, and the rapid reaction between iron and aluminium at this temperature leads to the formation of dross. Because of the high rate of oxidation of molten aluminium alumina may become entrapped on the steel surface on entry into the bath, preventing the deposition of a metal coating in these regions, and streaks of alumina may also contaminate the coating surface on removal of the work from the bath. For these reasons it is necessary to employ a molten fluoride flux bath operating at a similar temperature to the aluminising bath for pre-fluxing the steel; furthermore, a layer of molten flux must cover the aluminising bath to exclude air and to allow transfer of the fluxed steel direct to the aluminising bath without any intermediate exposure to the air.

A range of aluminium–iron intermetallic compounds of varying composition are rapidly formed when the steel is immersed in the molten aluminium, and the growth of the alloy layer is more rapid and continuous than in zinc during galvanising. The intermetallic compounds are harder and less ductile than the pure aluminium, and excessive alloying can lead to defective coatings. Control of alloying in the aluminising process is effected by the addition of 3–7 per cent of silicon to the aluminium; this slows down the rate of alloy formation and hence the thickness of the alloy layer in the coating, improves its uniformity and reduces its hardness.

As with galvanising, hot-dip aluminised coatings are normally specified by weight per unit area, commonly of the order of 150 g/m^2, which represents a coating thickness of about 25 μm. In service in a corrosive environment the oxide film naturally formed on the surface of the aluminium offers a high degree of corrosion resistance, so the coating corrosion is very slow. Even in highly polluted industrial atmospheres complete protection of the coated steel may be retained for periods well in excess of twelve years provided that the coating is sound and there are no areas of exposed substrate.

The potentials of iron and aluminium coupled in an electrolyte are not greatly separated and both can vary with the presence of films on their surfaces. Hence there is little, if any, sacrificial protection of steel by aluminium, and in some circumstances the steel itself may be initially anodic to the aluminium and so itself preferentially attacked. For these reasons large coating discontinuities cannot be tolerated in aluminised coatings because rusting of the steel substrate would occur, but it is notable that little, if any, rusting of the steel occurs at minute coating discontinuities or at cut edges of aluminised steel (probably because of stifling of the anodic reaction by adherent corrosion products).

A notable feature of aluminised coatings is their high degree of resistance to corrosion at elevated temperatures, making them very effective for use in equipment for handling hot flue gases. The reason for this excellent performance is the conversion at elevated temperatures of the pure aluminium coating into a thick, tenacious, inert film of alumina.

Hot-dipped tin on steel or copper (hot tinning)

Hot tinning also closely resembles galvanising and aluminising in treating acid-pickled and chloride-fluxed materials in molten metal. The operating temperature for the bath of molten tin is 300°C; the tinned work is withdrawn from it through a cover of palm oil of sufficient depth, and at a carefully controlled temperature, to ensure that the work finally emerges at a temperature of 240°C. If this 'controlled quenching' is omitted oxidation of the tin coating is likely to occur.

Although tin readily alloys with iron, the growth of the alloy layer during hot-tinning is slow and is limited to a thin layer adjacent to the steel. The alloy composition is simple, and these limitations are imposed by the very low solid solubility of iron in tin. In contrast, the solid solubility of copper in tin is much greater and hence thick, complex alloy layers are readily formed during hot-tinning of copper. The α-phase copper–tin alloy contains approximately 8 per cent tin, and the higher alloy phases likely to be present in the coating correspond to the compositions Cu_3Sn and Cu_6Sn_5.

Coating weights are usually in the range 25–75 g/m^2, a thickness of 1–5 μm, being limited in practice to these low levels because of the high cost of tin and also its nobility which ensures good corrosion resistance. Because of the nobility of tin it is cathodic to steel or copper in many corrosive media; this leads to severe localised pitting of the substrate at coating discontinuities. On the other hand, tin complexes form in many organic acids and in these conditions the coating may be anodic to steel — a property that is exploited, together with good solderability and non-toxicity, in the use of tinned steel for food canning.

Because of the thinness of tin coatings porosity is a major defect problem. The softness and ductility of tin enables some porosity to be eliminated by mechanical working after tinning, but porosity can be effectively reduced, and the appearance of the tin coating markedly improved, by means of the process known as flow brightening. In this process the tinned work is flash-resistance heated to remelt the coating momentarily, so that it flows evenly over the substrate surface and eliminates porosity.

Hot-dipped lead alloy on steel (terne coatings)

Molten lead does not wet the surface of most metals and hence simple immersion in pure lead does not produce complete and adherent coatings. If, however, a lead–tin alloy bath is used satisfactory coverage and adhesion can be obtained. With alloys containing 20–25 per cent of tin the resultant coatings are known as terne coatings, but lower tin contents (down to as little as 2 per cent) may be employed, and at these low levels the coatings are loosely described as lead coatings. The bath operating temperature varies with the percentage of alloying metal used.

Lead is a highly corrosion-resistant material, owing its protective action to the formation of insoluble corrosion products that stifle the corrosive reaction in most media except high-chloride environments. Resistance to the oxidising acids is particularly high. In addition, lead is very soft and extremely ductile and hence coated materials will withstand very extensive deformation without rupture of the coating.

Spray application

Sprayed metal coatings are obtained by making the coating metal molten and converting it into atomised globules in a spray gun. The molten globules are propelled to the surface to be coated at a speed of 100–150 m/s, are flattened on striking and adhere to it. It is not clearly understood whether freezing of the molten globules occurs on or before impact with the surface to be coated, but in view of the fact that the substrate material can be coated with only a very limited increase in its temperature it is likely that at least partial solidification occurs before impact. Coating thickness is built up by controlling the speed of movement of the spray gun and its distance relative to the surface to be coated, and by the application of a number of successive spraying passes.

The flattened solidified globules of coating metal adhere to the substrate surface purely by mechanical forces and there is no alloying action between the two metals. For this reason it is essential that the surface of the substrate shall be clean and of a sufficient degree of roughness to provide adequate mechanical keying between coating and substrate. This is achieved by carefully controlled grit-blasting immediately prior to metal spraying (as described in Chapter 2). As successive globules strike and flatten on the surface they become partially welded together and a cohesive coating is built up. Because of this method of growth the coating does not contain a defined crystalline microstructure; it contains fairly substantial percentages of oxides of the coating metal and also considerable porosity. Both the oxide content and the porosity can vary over quite a wide range according to the spraying process used and its method of operation. A typical sprayed zinc coating is shown in *Figure 3.3*

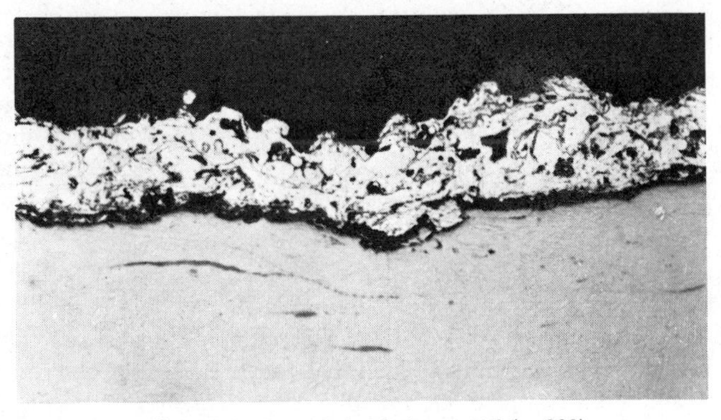

Figure 3.3 Sprayed zinc coating on steel (× 200)

Despite the presence of porosity and a high oxide content, the internal conductivity of sprayed metal coatings and also the conductivity across the coating/substrate interface are sufficiently good to enable the coatings to act either anodically or cathodically according to the choice of metals employed in the coating/substrate combination. However, apart from any electrochemical properties of sprayed coatings, their corrosion performance may differ from that of coatings produced by other means in that the porous nature of the coating allows some attack to take place within the coating thickness; corrosion products may then be readily entrapped, plugging the pores and stifling further corrosion.

Sprayed metal coatings are frequently used in combination with post-treatments in which greases, waxes, lacquers and inhibitor coatings are applied as pore sealants, and they also provide an extremely good base for final paint coatings. However, their corrosion protective role in their own right can be of a very high degree; indeed there is evidence that, in certain specialised cases, the application of sealer or paint coatings can reduce this protective role if the substrate is subsequently exposed by coating damage, since the effective available anode area will be greatly reduced by the post-treatment.[2] It is also possible, with certain combinations of coating and substrate metals, to apply heat-treatments after metal spraying in order to improve the corrosion resistance of the coating system. Such heat-treatments may induce diffusion alloying between the coating and the substrate or increase the proportion of the oxide of the coating metal in the coating itself. The alloy layers or metal oxides so produced may possess an intrinsically higher corrosion resistance than the coating metal as sprayed.

Because of the greatly reduced amount of heating of the substrate metal during metal spraying, there is minimal risk of distortion and damage to mechanical properties during the coating operation. Also, the high heat output of the spraying gun and the rapid cooling of the atomised globules allow the use as coatings of metals with higher melting points than the substrates to which they are applied. Add to these points the portability of the blasting and spraying equipment, high deposition rates and the ability to automate the process, and it is clear that metal spraying is a very versatile and useful process, which can be applied to a vast range of types and sizes of articles. Coatings may be applied in a factory at any convenient stage of component production, or *in situ* after construction has been completed. Further advantages of the metal spraying processes lie in the ability to obtain coatings of controlled thickness on shaped articles (although coatings can be more easily and economically obtained on simple shapes) and the ability to obtain selective coatings confined to chosen areas of a component by masking off during spraying.

On the debit side of the balance sheet, however, must be set the fact that metal spraying, taking into account the cost of grit blasting, is a more expensive method of coating than many other processes which are available and that during processing a consideraable quantity of the sprayed metal is wasted through "over-spray". The lower level of adhesion obtained with sprayed metal coatings must also be classed as a disadvantage but the higher porosity may be either a disadvantage or an advantage depending on the metals concerned and the

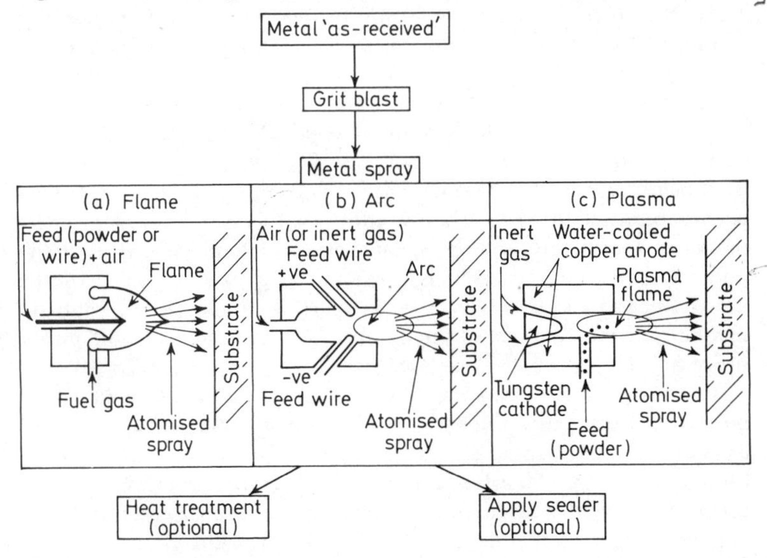

Figure 3.4 Flow-chart of application of sprayed metal coatings

service application. *Figure 3.4* shows a flow-chart of the application of sprayed metal coatings by the three basic methods, i.e. flame, arc and plasma.

Flame spraying

In flame spraying processes the coating metal is fed into the spray gun and melted by an oxyacetylene, oxyhydrogen or oxypropane flame. The molten metal is atomised by the action of a stream of compressed air — together with the streaming effect of the heating flame itself — and propelled from the gun nozzle toward the work to be coated.

The coating metal may be supplied to the spray gun in one of two different forms, wire or powder. With the wire process the coating metal is in the form of a wire of (usually) 2–3 mm diameter, which is fed to the central nozzle of the spray gun by means of powered rollers. As the wire issues from the nozzle into the centre of the heating flame its tip is continuously melted and atomised by the gas and air streams. With the powder process the coating metal is in the form of a fine powder contained in a hopper. Air or some other inert carrier gas is blown through the hopper and carries the powder particles suspended in the gas flow to the nozzle of the spray gun. The individual powder particles are melted in the flame and ejected in the gas and air streams.

There is little difference in the quality and properties of the coatings produced by the two methods of flame spraying, apart from a tendency toward slightly greater roughness in powder sprayed coatings. Coating porosity is normally in the range 10–15 per cent and the bond strength is of the order of 7 MN/m^2. The powder gun employs a larger flame than the wire gun and consequently the degree of heating of the workpiece tends to be somewhat greater. Because of this difference it is sometimes claimed that better adhesion can be achieved with the powder process, but wire-process operators counter-claim that the percentage of oxide present is greater in coatings produced by the powder process. In practice, both of these differences are minimal and can be varied by control of the operating technique used with either process.

The wire process can only be operated, obviously, for coating metals that can be produced in the form of wire of closely controlled diameter. The advantages of the process are continuity of operation limited only by the length of the wire coil, no risk of contamination of the coating metal, a more compact spray gun, and ease and rapidity of change of coating metal when desired. The powder process can be operated with any metal that can be produced in the form of a fine powder; it is possible, therefore, to produce coatings consisting of any desired composition of two or more individual metals (i.e. not limited by their ability to form alloys with each other) by the simple expedient of mixing the powders in the desired proportions either in the feed hopper or by using two separate hoppers and carrier-gas streams. Continuity of spray is limited by the size of the powder feed hopper, and practical considerations usually result in a smaller capacity than that available with the wire process. Contamination of the metal powder can occur if precautions are not taken to avoid this, and changing from one coating metal to another is more difficult because of the need to ensure that the hopper and pipes conveying the powder to the gun nozzle have been completely cleaned out. The particle size of the powder must be carefully controlled by sieving (the usual size limits being between 100 and 300 mesh) and dampness must be avoided to prevent clogging.

Arc spraying

In arc spraying the melting of the coating metal is accomplished by means of a constant voltage d.c. electric arc instead of a gas flame. In electric arc guns the coating metal is supplied in the form of two 2–3 mm diameter wires carrying the current supply, the arc being struck at their point of contact. As with flame spraying, the molten metal is atomised and ejected from the gun by means of a carrier gas, which is blown through the arc by means of a central nozzle immediately behind the two wire feeds.

In general, the advantages and limitations of the arc process are very similar to those of the wire flame spraying process. With replacement of oxyacetylene heating by electric current heating, however, there are some advantages of portability in the equipment. The higher temperature achieved in an electric arc enables the processing of coating metals of higher melting points than can be handled by flame spraying. Since all the heat required for melting is concentrated in the melting zone the heating of the substrate during spraying is less than with flame spraying techniques. The method is also claimed to produce coatings with a higher bond strength ($\sim$10 MN/m^2).

Plasma spraying

Plasma spraying resembles arc spraying in that a d.c. electric arc is employed for melting and atomising the feed metal, but in this case the arc is an ionised gas plasma struck between water-cooled metal electrodes that are not consumed in the process. In the plasma gun a water-cooled pointed tungsten cathode is mounted concentrically in the rear of a nozzle-shaped water-cooled copper anode. The carrier gas injected tangentially at the rear of the annular electrode gap is ionised to form the arc; the gas flow forces the arc forward into the restriction of the nozzle, where spiral flow produces a concentration of heat in the centre of the plasma arc. The exceedingly steep temperature gradient set up by this arc configuration produces a core temperature in excess of 20 000°C, whereas the temperature at the nozzle wall is as low as 250°C. The coating metal, in the form of powder, is carried in a second stream of gas and injected radially into the gun nozzle, so that the metal particles are melted and atomised during their passage through the plasma arc and ejected from the front of the nozzle by the gas stream.

The gas most commonly used in plasma guns is argon, though nitrogen may be used to reduce the costs of processing. In order to increase the core temperature for spraying more refractory materials a small percentage of hydrogen can be incorporated or, where hydrogen embrittlement might be a problem, helium can be used instead.

The major advantages of plasma spraying (apart from the increased range of refractory materials that can be sprayed) lie in reduced coating porosity and increased adhesion to the substrate, coupled with very limited heating of the substrate material. Porosity values in the range 1–10 per cent can be readily achieved and adhesion is typically of the order of 30 MN/m^2. The disadvantage is economic, in that the plasma process is fairly expensive by comparison with flame and metal arc spraying processes.

Applications of sprayed metal coatings

The range of coating metals that can be sprayed and substrates that can be coated is almost limitless; coating thicknesses employed may vary, according to the application, from some tens of micrometres for protective coatings that are to be subsequently painted up to several millimetres for coatings that offer extremely high corrosion and wear resistance and hardness.

Zinc and aluminium and their alloys are used as sprayed coatings for the protection of steel against atmospheric corrosion, in thicknesses in the range 50–150 μm, while somewhat thicker coatings are used for immersion in natural or sea waters. These coatings provide sacrificial protection to the steel substrate in the same manner as their respective hot-dipped counterparts, though no element of alloying with the substrate enters into the corrosion reaction. The stifling action of corrosion products is greater than with hot-dipped or electrodeposited coatings on account of the porous nature of the sprayed coatings, and consequently somewhat longer lives may be obtained in service. In the case of aluminium sprayed steel some diffusion alloying can be achieved by post-spraying annealing treatments; because of this alloying and the increased percentage of inert aluminium oxide present in the annealed coatings, a very high degree of resistance to elevated temperature corrosion can be obtained.

Aluminium, zinc and their alloys can also be used very successfully as sprayed metal coatings for the protection of high-strength aluminium alloys of the aluminium–copper–magnesium and aluminium–zinc–magnesium types against stress corrosion and exfoliation corrosion. Failure of these alloys from these causes can be very rapid in service, but the use of a sprayed metal coating of the order of 125 μm thickness provides complete protection for periods in excess of ten years, and the coatings protect sacrificially at gaps exposing the substrate metal.

Sprayed zinc or aluminium coatings on steel are of special value in applications where friction grip bolting is involved. Slip factors of the order of 0.45–0.55 are readily obtained with sprayed zinc coatings and in the case of sprayed aluminium coatings the slip factor can rise as high as 0.7.

Materials highly resistant to aggressive acidic environments or to high-temperature oxidation (such as stainless steels and alloys of copper, nickel and chromium) are commonly employed, often with post-spraying heat-treatments and/or grinding or polishing, to improve wear resistance and bearing properties.

Sprayed tin coatings can be used, but because of the nobility of tin the pores in the coatings must be sealed by flow brightening to avoid preferential attack on the substrate metal.

Lead may be metal sprayed on to steel for coatings resistant to acidic, sulphurous gases, protection being obtained by the production of lead sulphate in the pores of the coating, which stifles the corrosive reaction. Here again the coating metal is cathodic to the substrate steel, and precautions must be taken in service to avoid mechanical damage to the coating since the steel will be preferentially attacked if exposed to the environment.

Refractory metal coatings and ceramic coatings are applied by metal spraying for corrosion protection in very high temperature applications such as furnaces, burners, turbines and jet engines.

A specialised application of sprayed metal coatings is in combating corrosion fatigue and fretting corrosion. The fatigue resistance of metals, particularly non-ferrous metals, can be improved by the incorporation of compressive stresses in the surface layers. Grit blasting a metal surface prior to metal spraying compressively stresses its surface and this can improve fatigue life; the presence also of a suitable protective sprayed metal coating can combat the corrosive factors in applications where corrosion-fatigue conditions apply. In fretting corrosion the oxygen concentration cell formed in a bolted crevice, together with finely divided metal powder produced by abrasion during small-amplitude movement of the components of the joint, cause localised corrosion; a sprayed metal coating offers higher frictional resistance to relative movement and a protective coating, both of which can contribute to a reduction in fretting corrosion.

Chemical deposition

Probably the simplest example of chemical deposition is the ability of copper to plate out on to iron immersed in a solution of a copper salt

— usually copper sulphate. The process is one of simple substitution of iron ions in solution in place of copper ions. However, the process is not of great practical use in metal-coating technology since deposition stops when the iron is completely covered by the copper, so the deposit remains extremely thin; the coating is also porous and only very poorly adherent to the substrate. Two fields in which copper immersion coatings are used are the decorative copper colouring of iron and steel articles, and the coppering of steel sheet or wire for temporary protection and to promote easy lubrication during deep-drawing and pressing operations. Immersion tinning of copper alloys in solutions of stannous salts is used in soldering applications, and zinc is deposited on aluminium by immersion in hot, alkaline zincate solutions in order to provide a thin coating as a basis for subsequent electroplating with other metals — notably copper, nickel and chromium. Both tin and silver coatings can also be obtained by chemical immersion treatments for purely decorative finishing.

A second type of chemical deposition is achieved autocatalytically, whereby the coating metal deposits on a metallic or metal-activated surface and the coating thickens with a more or less linear growth rate as long as the compositional balance of the solution is maintained. Solutions of this type are commonly known as 'electroless plating' solutions. Metals that can be autocatalytically plated are copper, nickel, iron, cobalt, silver, gold, platinum and palladium. Of these, copper and nickel are probably the most widely used (in engineering and electronic applications or for the metallising of plastics materials in preparation for electroplating), and there is a somewhat more limited usage of silver and gold in certain electronic applications.

Electroless plating solutions for copper and nickel consist of aqueous solutions of salts of the respective metals. Alkaline solutions are used for both copper and nickel, and acid solutions may also be used for nickel. Essential requirements of these solutions are suitable reducing and buffering agents; stabilisers and accelerators may also be incorporated in the formulations. The reducing agent normally employed in electroless copper baths is formaldehyde or hydrazine; hypophosphites and borohydrides are used in electroless nickel baths. Electroless copper baths are usually operated at or a little above room temperature; they tend to be somewhat unstable, and consequently the addition of the reducing agent is frequently not made until the solution is required to be worked, after which the bath is worked to exhaustion and discarded. Electroless nickel baths, however, which operate at higher temperatures (60–100°C), are much more stable; they can be stored, fully formulated, at room temperature and may be operated over long periods with the addition of suitable replenishment chemicals.

In operation, the metal to be coated acts as a catalyst to allow the reducing agent to reduce the copper or nickel ions so that the metal is deposited with evolution of hydrogen. A simplified equation for the overall reaction is

$$(Ni, Cu)^{++} + 2(RA) + 2OH^{-} \rightarrow (Ni, Cu)^{\circ} + 2(RAO) + H_2$$

where (RA) is the reducing agent and (RAO) is its oxidation product.

Copper is deposited as relatively pure metal (with perhaps some cuprous oxide incorporated) at a rate of 2–12 μm/hour; nickel co-deposits with either phosphorus or boron (depending on the reducing agent used in the solution) at a rate of 12–30 μm/hour.

Electroless coating processes are expensive to operate but have the advantage that the deposit thickness is completely even irrespective of the complexity of the surface geometry of the article being coated. In the case of electroless nickel coatings the incorporation of phosphorus or boron in the deposit increases hardness and brittleness, and alters the corrosion resistance compared with that of the (purer) electrodeposited metal. These properties of the nickel deposits can also be modified by subsequent heat-treatment. Adhesion of the deposits is dependent upon chemical bonding, assisted by mechanical keying to a roughened surface, and there is no alloying with the substrate metal unless diffusion is induced by heat-treatment after electroless plating.

Pretreatment prior to electroless plating must be carefully carried out and varies with the substrate material. Steels should be electrolytically cleaned and acid etched to micro-roughen the surface. Copper alloys for electroless nickel plating must be thoroughly cleaned and etched and, since nickel will not reduce directly on to a copper surface, should have their surfaces catalysed with palladium chloride prior to treatment in the electroless bath; thorough rinsing off of excess palladium chloride prior to immersion must be employed. Aluminium alloys can be electroless nickel plated after only pickling and etching, but more effective results are achieved if a further pretreatment is employed to deposit a zinc coating by zincate immersion prior to electroless nickel plating.

In the case of non-metallic substrates (such as plastics) it is essential first to convert the surface of the non-metallic material from the hydrophobic (i.e. water-repellent) condition to the hydrophilic (i.e. water-receptive) condition and to micro-roughen the surface by solvent and/or acid-etching processes. The surface must then be catalysed with palladium from a palladium chloride solution and thoroughly rinsed before electroless plating with copper or nickel.

Electrodeposition

Coatings are obtained by electrodeposition on to a conducting substrate. The metal to be coated is immersed in a conducting solution containing a salt of the coating metal and is made the cathode by applying an e.m.f. from an external source. The anode in the cell may consist of a rod or sheet of the coating metal — in which case it passes into solution as deposition takes place on the cathode, so maintaining the metal ion concentration in the solution. Alternatively, an anode of an inert material may be used — in which case the metal ion concentration must be maintained in the solution by suitable additions of metal salts as electrolysis proceeds. See *Figure 3.5.*

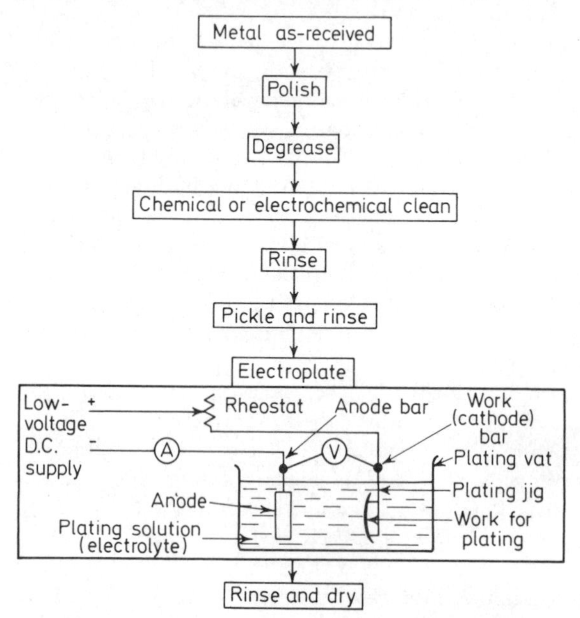

Figure 3.5 Flow-chart of application of metal coatings by electrodeposition

The process of electrodeposition is, essentially, the cathodic side of the same electrochemical reaction as that which causes corrosion on the anodic side (see Chapter 1), the reaction being carried out under controlled conditions of electrolyte composition, potential and current density selected to favour the cathodic reduction of metal ions, so metal is deposited rather than anodically oxidised to form metal cations or other oxidised species.

Coating deposition initiates by nucleation at defects in the crystal lattice of the substrate metal, such as dislocations at the surface, with subsequent crystal growth of the deposited metal from the nucleated sites. By this mode of growth an adherent crystalline metal coating is built up on the substrate, bonded to it by atomic linkages, which ensures complete adhesion, and without the growth of alloy layers between coating and substrate (unless the metals concerned are such that diffusion of one into the other can occur in the solid state at the temperatures applicable during deposition or in subsequent storage or use).

Electrodeposits may be of pure metals, mixed metals, alloys or metals mixed with non-metallics. Pure metal deposits are obtained from electrolytes containing salts of that metal; mixed metal and alloy deposits require electrolytes containing salts of the component metals, which either co-deposit independently or deposit directly in the alloy form; non-metallics are incorporated into electrodeposits by using solutions in which the non-metallics are present either as simple suspensions or in the form of complexes that are broken down at the cathode.

In aqueous solutions the reduction of hydrogen ions and water to hydrogen gas is a possible alternative to reduction of metal cations or anions to deposit metal, and the more negative the potential of the M^{z+}/M system the greater the tendency for hydrogen evolution to occur. Zinc (or possibly manganese) is the most negative metal that can be deposited in practice from aqueous solutions; in the case of the more negative metals non-aqueous solutions or fused salts must be used. The aluminium can be electrodeposited from a solution of $AlCl_3$ and LiCl in anhydrous ether, and titanium can be plated from fused salts.

Properties of electrodeposits

During the early stages of deposit growth defects may be present in the crystal structure, so pores may occur in the coating. However, provided that the operating parameters of the electrodeposition process have been chosen and controlled so as to produce the best growth characteristics in the deposit, these defects will be eliminated as the thickness increases and, in general, coatings will be pore-free when their thickness reaches a few micrometres. There are, however, occasions when defects in the structure of a coating may be deliberately induced for special purposes (micro-discontinuous chromium — see Chapter 4).

Since the electrodeposition of metals takes place in accordance with Faraday's Law ('the mass of metal decomposed by electrolysis is directly proportional to the quantity of electricity passed through the solution and proportional to the chemical equivalent of the metal'), it follows that the average thickness of an electroplated coating of a given metal can be quickly and easily calculated from a knowledge of the current, the plating time, the surface area of the plated article and the chemical equivalent of the plated metal. However, there is one important additional factor that affects the calculation, namely the cathode efficiency of the plating operation. The simple calculation applies only for metals, such as copper, that have approximately 100 per cent cathode efficiency when deposited from an acid copper sulphate solution, but the efficiency varies enormously from one plating solution and one metal to another; it can be as low as 8–18 per cent for chromium, i.e. only one tenth to one fifth of the theoretically possible thickness will be deposited in a given time. The weights of electrodeposited metal theoretically produced by one ampere-hour of electric current are listed in *Table 3.1*.

Table 3.1 CALCULATED ELECTRODEPOSITION RATES FOR METALS*

Metal	*Atomic weight*	*Valency*	*Chemical equivalent*	*Deposition rate at 100% efficiency* g/Ah
Cadmium	112.40	2	56.20	2.10
Chromium	52.00	2	26.00	0.97
		3	17.33	0.65
		6	8.67	0.32
Cobalt	58.93	2	29.47	1.10
Copper	63.54	1	63.54	2.37
		2	31.77	1.19
Gold	196.97	1	196.97	7.36
		3	65.66	2.45
Iron	55.85	2	27.92	1.04
		3	18.61	0.70
Lead	207.19	2	103.60	3.87
		4	51.80	1.93
Nickel	58.71	2	29.35	1.10
Platinum	195.09	2	97.55	3.64
		4	48.78	1.82
Silver	107.87	1	107.87	4.03
Tin	118.69	2	59.35	2.21
		4	29.67	1.11
Zinc	65.37	2	32.67	1.22

*Based on a table in *Canning Handbook on Electroplating*, 21st edn (1970)

A further point to be considered concerning the thickness of electrodeposited metal is that the deposit thickness varies with variations in the distance between anode and cathode. The ability of a plating solution to overcome these variations is known as its 'throwing power' (or, more properly, its macro-throwing power); this property varies from one metal to another and may also be influenced by bath composition and operating variables. Copper is a good example of a metal with good throwing power, and chromium is a metal with poor throwing power. Because of throwing power limitations deposit thickness varies with the geometry of the surface of the article being plated, so thickness builds up on sharp edges and general convexities and can be seriously depleted in sharp concavities and in angled recesses.

The ability of a plating solution to reduce the degree of surface roughness of the substrate, i.e. its micro-throwing power as opposed to its macro-throwing power, is an entirely separate property known as 'levelling'. An electrolyte with good levelling properties produces a deposit that becomes progressively smoother than the substrate as its thickness builds up. It is thought that polarisation differences between micro-peaks and micro-valleys of the substrate surface affect ionic diffusion to and/or the rate of adsorption on the surface, so locally changing the rate of deposition. Levelling properties are usually controlled by the incorporation of special addition agents to the formulation of the plating bath; usually these addition agents are organic compounds — e.g. coumarin in nickel plating solutions. Good levelling and good throwing power are often co-properties of a plating solution, but this is by no means always the case — e.g. zinc has good throwing power but very poor levelling properties.

When electrodeposition is carried out using simple solutions of the metal salts the deposits obtained are usually of matt appearance. In such cases a bright finish can only be obtained by polishing or buffing after plating — a costly and time-consuming operation. However, bright deposits can often be obtained direct from the plating bath by incorporating specific addition agents in the electrolyte composition. Surfactants and colloids are normally used for this purpose, acting through their ability to complex metal ions and their effect on adsorption and localised cathodic polarisation. They may also influence the form of crystallisation of the electrodeposits (as, for example, the lamellar micro-structure of bright nickel deposits compared with the columnar micro-structure of dull nickel deposits). Bright deposits are obtained only over a limited range of plating current densities (again modifiable by the use of specific addition agents) so dullness may be encountered in practice on the edges of shaped articles where high current densities occur during plating.

The internal stress of an electrodeposit is a property of prime importance to its use for corrosion-resistant purposes. In general, corrosion resistance is reduced by an increase in internal stress, owing to the increased liability of the deposit to fracture as corrosion develops, thus exposing the substrate and destroying the protective properties of the coating. Internal stress in deposits may be caused by the degree of lattice misfit between the substrate and the initial atomic deposit layers, or by the mode of deposition and crystallisation of the metal from the electrolyte. Lattice misfits cannot, of course, be influenced by plating conditions; internal stress due to this cause can only be reduced by the incorporation of an additional, different, coating between the substrate and the chosen metal so that the lattice difference is apportioned between the two interfaces. Stress due to the mode of electrodeposition and crystallisation can often, however, be markedly modified by changing the electrolyte composition or the plating parameters. Taking as an example, again, electrodeposited nickel, the dull deposits are low in internal stress while the bright deposits are more highly stressed.

A feature of the electrolytic process is the production of hydrogen at the cathode. Hydrogen molecules produced by reduction of hydrogen ions or water molecules may be liberated in the gaseous state, and hydrogen in the atomic form may be absorbed into either the coating or the substrate metal. The extent of any or all of these reactions may be influenced by the conditions under which electrolysis (i.e. electroplating, in this context) is carried out, since the extent of hydrogen evolution is inversely proportional to the cathode efficiency of the plating process.

If hydrogen is liberated as gas bubbles at the cathode the presence of these bubbles can interfere with the deposition process and cause bare spots, irregular deposits or deposits with crystal defects. In most commercial plating operations cathodic hydrogen evolution is kept to a minimum, but if complete suppression is not possible removal of the gas from the cathode surface is encouraged by agitation of either solution or cathode.

If hydrogen produced at the cathode is absorbed into the coating or substrate metals embrittlement can occur (as, for example, in the case of high-tensile steels during plating with zinc or cadmium) and in such cases post-plating heat-treatments are usually specified in order to diffuse away the hydrogen and so prevent cracking in service.

Design for electroplating

Because of the nature of the electrodeposition process, the mode of growth and the other properties of the deposits, it is important to consider carefully the design of components that are to be electroplated so as to ensure the best results.

The need to provide efficient contact points for supplying electric current to the work and the fact that the coating will be broken at the point where the jig contact is subsequently removed necessitates careful positioning of the contact points in such places that the appearance and performance of the coating are not affected. The shape of a component should be kept as simple as possible in the interests of uniformity of deposit thickness, and sharp angles and recesses should be avoided. Where complex shapes must be employed it is possible to overcome thickness variations to some extent (and at

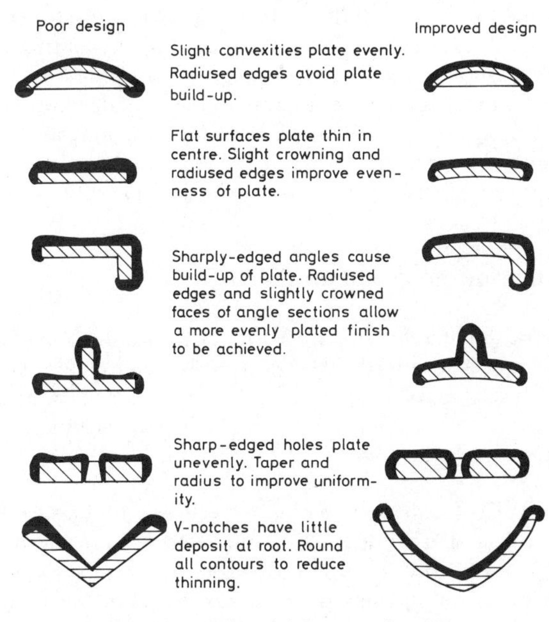

Figure 3.6 Effect of design upon plateability

high additional cost) by using supplementary anodes or anodes of conforming shape during electroplating. Also, since deposition occurs freely only on surfaces that are opposite an anode, the back surface of a component can be coated only by suspending it midway between two parallel rows of anodes. Similarly, the internal bores of tubular sections can be electroplated only by the use of central, internal, auxiliary anodes. Some features of design that affect plateability are shown in *Figure 3.6*.

Limitations on the size of article that can be successfully electroplated arise from the dimensions of the largest available electrolyte

tanks and the capacity of mechanical handling equipment. However, in the case of sheet or strip, electrodeposition can be carried out on a continuous process, feeding the work into and out of the processing tanks and supplying the electrical contact by means of rollers. At the other end of the scale, small components such as fasteners, auxiliary fittings etc. that would be impossible or uneconomic to wire up on jigs can be electroplated by treating in perforated barrels immersed in the electrolyte. The cathodic connection is made by mass contact through the load of components in the barrel, which is continuously rotated during the plating operation so that deposition occurs evenly on all the components as they shift position relative to each other in the load. The process is slower in building up a given coating thickness than is the case with jig plating, since deposition on any individual component only occurs when it shifts to the outside of the load adjacent to the periphery of the barrel. Some coating may also be lost by bi-polar effects within the mass of the load and, possibly, also by mechanical abrasion and by simple chemical re-solution.

Preparation for electroplating

The preparation of a metal surface for electroplating must be carried out to the highest standards in order to ensure complete coverage and good-quality adherent coatings. The metal surface must be completely free from mechanical defects that will mar the finished surface, free from scale, oxide films or greases, and completely chemically clean.

The presence of scales, oxide films or greases inhibits electrodeposition in the affected areas, leading to uncoated portions or, if the areas are small enough to allow the coating to grow over them, regions where the adhesion between the coating and the substrate is inferior and flaking or stripping may subsequently occur.

The smoothness of the surface must be controlled (to the degree necessary to achieve the appearance required on the finished article) by means of abrading, polishing and buffing operations. Surface laps, pores or fissures must be avoided since they may entrap processing solutions, which may subsequently react chemically with the substrate and/or the coating metals and lead to localised failures.

It follows from the above that combinations of any or all of the methods of pretreatment described in Chapter 2 (abrading, polishing, degreasing, cleaning, etching and attendant rinsing) must be employed as appropriate to particular applications, so as to present for electroplating a completely clean metal surface of the requisite quality.

Post-treatment of electrodeposits

On completion of the process of electrodeposition it is essential to employ complete and thorough rinsing to remove all processing solutions. Plating solutions are usually strongly acidic or strongly alkaline, and any solutions that are not promptly removed from the metal surface when the current supply is disconnected are likely to attack the surface, causing pitting and/or staining. After thorough rinsing the plated articles should be rapidly dried to prevent water staining or general corrosion by retained miosture.

In many cases (e.g. zinc and cadmium plating) the plated metal surface may be further treated by chemical passivation to retard the onset of corrosion in mildly corrosive conditions. Alternatively (e.g. with copper plating) a clear lacquer may be applied to the plated surface so as to prevent atmospheric oxidation tarnishing.

Coating metals applied by electrodeposition

The range of metals that can be electrodeposited is extremely wide, indeed there are few metals that cannot be plated (although organic solutions or fused salts must be used for very electronegative metals). In some cases electrodeposition would have little practical application, in others plating is mainly for engineering or electroforming applications rather than corrosion prevention. Coating metals used specifically for corrosion control include the following.

Cadmium

Cadmium is directly plated on to iron and steel to provide a sacrificial protective coating similar in action to that of a zinc coating. It may also be used, in conjunction with a tin undercoat, as a coating on copper alloys. Coating thicknesses are usually up to a maximum of 25 μm; heavier deposits are generally ruled out by the high cost of cadmium metal. Thinner deposits (~2.5 μm) may be used as undercoats for zinc on cast iron.

The metal is deposited principally from cyanide solutions (although fluoborates have been used), and cadmium metal anodes are used. Deposition is usually at temperatures of 20–35°C, the efficiency is 90–95 per cent, and the throwing power is good. Deposits from the simple bath are dull, but bright coatings can be obtained from baths containing addition agents.

Chromium

Chromium is very widely used as a decorative electrodeposit, because of its high lustre, in the form of coatings 0.3–1.0 μm thick that form topcoats over protective deposits of metals such as nickel. The metal itself is virtually inert in most commonly encountered corrosive environments and, because it is very hard, thicker deposits are used directly deposited on to steel or other substrates for wear-resistant applications.

Electrodeposition of chromium is almost universally from a chromic acid/sulphuric acid based solution, using lead anodes and an operating temperature in the range 37–65°C depending upon the type of plating solution; the chromium is replenished periodically, to replace that deposited, by additions of chromic acid. Deposits from these baths are fully bright but the throwing power is poor, leading to uneven deposit thickness and incomplete coverage in recessed areas of plated articles; furthermore, the cathode efficiency is low — in the range 8–18 per cent according to the type of solution and the operating conditions. Higher cathode efficiencies can be obtained from baths catalysed with silicofluoride (up to 25 per cent efficiency) and from the tetrachromate (Bornhauser type) baths (up to 30 per cent efficiency). Considerable research and development is being carried out into the process of depositing chromium from the trivalent chromic chloride based solutions, which offer increased cathode efficiency and speed of deposition together with lower operating temperatures. The chromium deposits obtained from trivalent baths are slightly darker in hue than deposits obtained from the hexavalent chromic/sulphuric baths.

Thin chromium deposits always contain minute discontinuities that allow penetration of corrodents to the underlying metals, resulting in localised corrosion. If attempts are made to eliminate these discontinuities by increasing the thickness of the chromium deposit, the highly stressed nature of the chromium leads to unsightly macro-cracking, as shown in *Figure 3.7.* Modification of the formulation of the chromic-acid plating solution and the conditions of electrodeposition allow some increase to be made in the thickness of deposits (up to approximately 1 μm) without the occurrence of macro-cracking, and hence a reduction in the number of discontinuities through which corrosion can occur. This type of deposit is known as 'crack-free chromium'; the deposit continues to have a high internal stress, however, and spontaneous cracking can occur, particularly in those regions where the deposits are thickest, if the plated components are flexed in service. In addition, severe macro-cracking occurs in regions of shaped components where high-current-

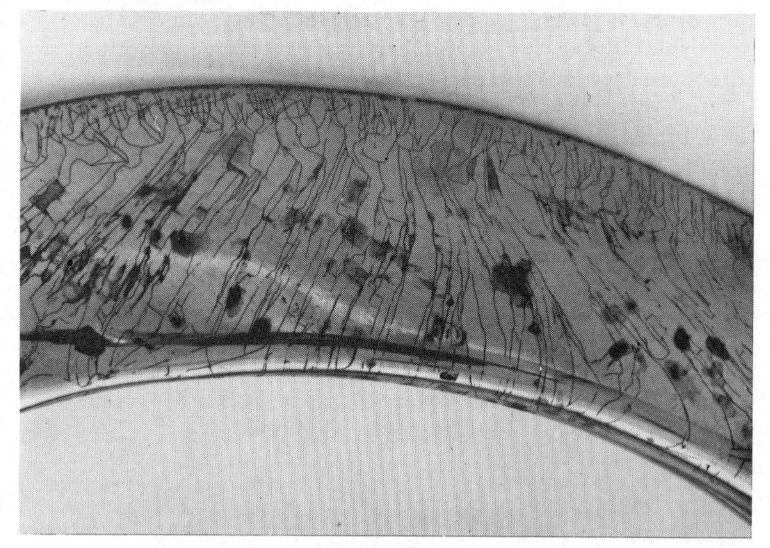

Figure 3.7 Macro-cracking of chromium electrodeposit (× ½)

density plating conditions cause excessive build-up of deposit thickness so that the 1 μm maximum is exceeded.

An alternative approach to the problem of overcoming the localised corrosion due to discontinuities in the chromium deposit is to reduce

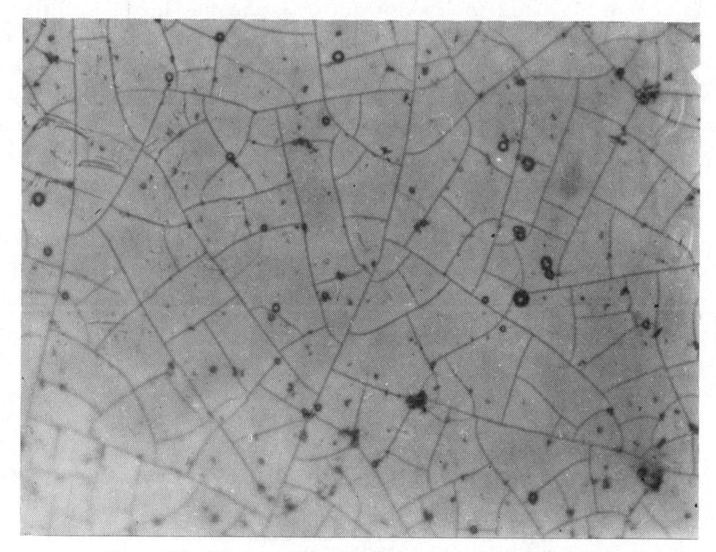

Figure 3.8 Micro-cracking of chromium electrodeposit (× 1000)

the corrosion current density at each individual anodic corrosion site by producing a vast number of micro-discontinuities in the chromium deposit. When this is done, corrosion at any given point of the underlying metal is slowed down and the period of protection of the substrate correspondingly increased. This process is achieved by modifying the chromium deposition process so as to induce the formation of a pattern of fine micro-cracks, having a density of more than 250 cracks per linear centimetre, the cracks themselves being invisible to the naked eye. The micro-cracks can be observed under the microscope as shown in *Figure 3.8*. Micro-cracked chromium is produced either by the use of special addition agents to the chromium plating bath or by depositing a normal decorative chromium on top of a thin layer of highly stressed nickel so that spontaneous micro-cracking of the chromium occurs. Chromium deposited from trivalent baths is normally in the micro-discontinuous condition.

Copper and its alloys

Electroplated copper is employed for four main purposes:

(a) As an undercoat for various protective coating systems such as nickel + chromium or precious metal deposits. In these applications the ability of the copper deposits to enhance brightness and to smooth out irregularities in the substrate through its good levelling properties is exploited as a means of reducing the amount of preliminary polishing required for decorative finishes. It may also be necessary as an undercoat in cases where the other deposits (e.g. nickel) cannot readily be deposited directly on to the substrate metal (e.g. zinc alloy diecastings or zincate-treated aluminium).

(b) As a protective coating for steel that requires subsequent processing (e.g. hardening).

(c) As a coating protective in its own right against moderately corrosive conditions (e.g. copper-plated steel for domestic and office fittings), although there will usually be further protection by lacquers or colour finishes such as 'oxidised' or 'sulphided' treatments.

(d) For electroforming processes such as in the manufacture of gramophone records or printing plates.

Three basic types of electroplating bath are used for copper plating:

(a) Alkaline copper cyanide solutions operated at around 65–70°C and having cathode efficiencies ranging between 50 and 100 per cent.

(b) Acid copper sulphate solutions used at 20–50°C and having approximately 100 per cent cathode efficiency.

(c) Alkaline copper pyrophosphate solutions used at about 55°C and having approximately 100 per cent cathode efficiency.

In addition to these three types of solutions fluoborate and sulphamate solutions may be used, mainly for electroforming applications. Copper anodes are used in all copper plating solutions.

The cyanide baths can only be used for thin deposits, and are frequently employed for initial 'strike' deposits approximately 1 μm thick on steel or zinc prior to the deposition of heavier deposits from the acid bath, which cannot be used directly on these substrates. The other copper plating solutions all have good throwing power and good levelling properties, the latter property being improved by the use of organic addition agents, which also enhance the brightness of the deposits.

Copper alloy electrodeposits can be obtained by plating from complex alkaline cyanide solutions at temperatures in the range 30–90°C according to the solution used. A range of brasses or bronzes may be plated using anodes of the appropriate alloy composition, the cathode efficiency and the exact composition of the electrodeposits varying with the current density used in the deposition process. Most of the deposits can be obtained in a reasonably bright condition, but levelling properties are generally poor or non-existent. Thin deposits are normally applied to steel for decorative applications, either alone or in combination with nickel undercoats for improved levelling, and usually with lacquer finishes to resist atmospheric tarnishing. In some applications a decorative chromium topcoat may be employed, but the copper alloy deposits are frequently highly stressed and this can lead to serious cracking of the chromium. Bronze electrodeposits may also be used as protective undercoats for nickel and/or chromium deposits in applications involving wear-resistance in highly corrosive environments (e.g. hydraulic mining equipment).

Gold and its alloys

Deposits of gold and its alloys are used for electrical and electronic applications, for high quality decorative finishes or as high-temperature oxidation-resistant coatings. Coatings may range in thickness from as little as 0.05 μm for decorative finishes to as much as 30μm in specialised electrical and electronic applications, although owing to the cost of the metal the majority lie at the bottom end of this thickness range.

Electrodeposition is most commonly from the cyanide baths

(alkaline, neutral or acid), although the acid chloride bath can be used. Addition agents may be incorporated to improve brightness. Anodes may be of gold or of an inert material such as graphite or stainless steel.

As with chromium electrodeposits, thin deposits of gold are prone to porosity, which can adversely affect their ability to protect the substrate. Close control of plating conditions is required to minimise porosity, and careful attention must be given to the quality of preparation of the substrate for plating. Thin undercoat deposits may also be used with gold electrodeposits to provide additional protection to the substrate.

Lead and its alloys

Lead has a very high resistance to corrosion in acidic environments, and electrodeposits obtained from acidic fluoborate, fluosilicate or sulphamate solutions are used for the protection of ferrous materials or copper-based alloys. Electrolysis is at about 40°C using lead anodes, and the cathode efficiency is 100 per cent. Coating thicknesses in common use are in the range 10–100 μm, the heavier deposits being normally used for chemical plant applications.

If the lead fluoborate plating solution is modified by adding tin salts and substituting lead–tin alloy anodes for pure lead anodes, deposits of lead–tin alloys can be obtained covering a range of alloy compositions dependent on the solution formulation and the anode composition. Similarly, by adding antimony salts as well as tin salts to the solution, ternary alloy deposits can be obtained. These ternary alloy deposits are used for bearings and in electronic applications where soldering processes are involved.

Nickel and its alloys

Nickel is one of the principal metals used as electrodeposits for corrosion control; a wide range of processes exist, offering coatings with different physical, mechanical and corrosion properties. Most of the solutions used are based on the Watts nickel bath consisting of mixed nickel sulphate and chloride salts, although baths based on nickel chloride alone or on nickel sulphamate are also used. Electrodeposition is at 40–70°C using pure nickel anodes and the cathode efficiency exceeds 95 per cent.

Deposits obtained from the Watts bath or from the simple chloride bath are dull and require a considerable amount of mechanical polishing to produce the brightness required for decorative applications. In order to overcome this disadvantage solutions were developed for depositing nickel in the bright condition. Baths containing cobalt sulphate produce bright deposits with good ductility, but there is little or no levelling action during plating. The most widely used bright-nickel plating solutions employ organic addition agents to promote brightness and levelling; fully-bright and levelled deposits can be readily obtained and the solutions have good throwing power. In general, the bright nickels have lower ductility and higher internal stress, but these disadvantages are reduced if the sulphamate bath is used; this bath can be operated at higher current densities, giving more rapid deposition, but at higher cost.

An adverse feature of the bright nickels with respect to corrosion control is their lower corrosion resistance, owing to the presence of sulphur incorporated in the deposits from the organic addition agents in the baths. Semi-bright nickel baths have, therefore, also been developed in which the addition agents impart high degrees of levelling but only partial brightening, and in which the sulphur content of the deposits does not exceed 0.005 per cent. These deposits are more corrosion-resistant than the organic bright nickels and in practical corrosion control applications the two types of deposits are used in combination to produce the duplex nickel coatings consisting of two- or three-layer bright and semi-bright composites (see Chapter 1).

Nickel coatings for corrosion protection range from 5 μm to 40 μm in decorative applications used alone or in combination with chromium overlays, according to the nature of the substrate (steel, zinc alloy, copper or copper alloys, aluminium or aluminium alloys or plastics materials) and to the severity of the corrosive environment encountered. Thicker deposits may, of course, be employed for special corrosive applications such as in chemical plant or for electroforming.

Electrodeposition processes have been developed in which minute inert, insoluble particles are incorporated in suspension in the plating bath. When these solutions are used nickel deposits may be produced having a matt or satin finish. Alternatively, by limiting the deposition from the modified bath to thicknesses of 1–2 μm and depositing on to the surface of a bright nickel deposit the finished coating retains a bright appearance, but thin chromium deposits applied over the modified nickel layer have a large number of micro-pores (more than 10 000 per cm^2) produced because the chromium does not deposit on the surface of the individual non-conducting particles. The corrosion resistance of the complete coating system (known as micro-porous chromium) is considerably greater than that of systems with the normal decorative chromium (see Chapter 1).

The mechanism of corrosion protection of the different types of nickel + chromium coating systems is as follows. In the system shown in *Figure 3.9(a)*, corrosion at a defect in the conventional chromium layer rapidly attacks the underlying bright nickel owing to the high corrosion current density (small anode/large cathode), causing undercutting and accelerated attack on the substrate when the nickel layer is penetrated. In the system shown in *Figure 3.9(b)*, since bright nickel corrodes more rapidly than semi-bright nickel the corrosion pit widens laterally in the bright nickel layer and penetration of the semi-bright nickel layer is delayed, with consequent enhanced protection of

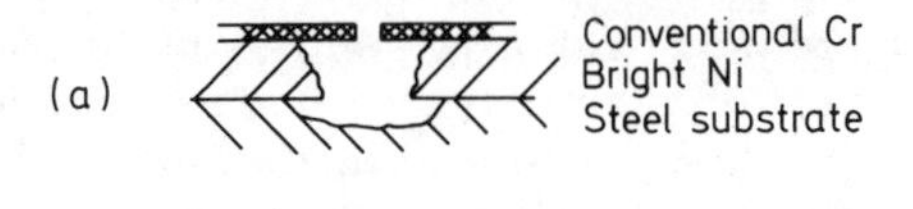

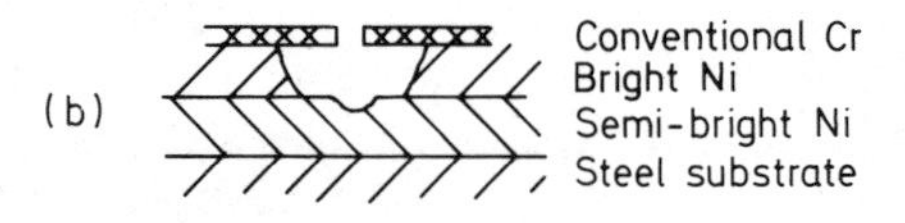

Figure 3.9 Mechanism of corrosion protection of different types of nickel + chromium coating

the substrate. Increasing the number of discontinuities in the chromium layer by the use of micro-porous or micro-cracked chromium – *Figure 3.9(c)* – increases the anode/cathode ratio with the underlying nickel. As a result the corrosion current density at each corrosion site in the bright nickel layer is reduced. Penetration of the bright nickel layer is thus markedly reduced and the overall protective life of the system is enhanced.

Special precious-metal electrodeposits

Platinum, rhodium and ruthenium are all electrodeposited for special applications, which include high-quality decorative finishes and electrical and electronic applications. Because of high cost (and, in the case of rhodium and ruthenium, the highly stressed nature of the electrodeposits, which causes spontaneous cracking) deposit thicknesses are limited to a few micrometres. The nobility of all three metals makes them highly corrosion resistant as coating materials.

Tin and its alloys

Tin may be electrodeposited from three different types of solution.

(a) Alkaline stannate solutions are used for normal batch jig plating, being operated at temperatures of the order of 65°C with either pure tin or insoluble nickel-plated anodes. These solutions have high cathode efficiency (60–90 per cent) and excellent throwing power but the deposits obtained are not fully bright.

(b) Acid fluoborate solutions also have good efficiency and throwing power, and with the incorporation of addition agents produce fully-bright deposits.

(c) Acid sulphate solutions have very poor throwing power, though bright deposits can be obtained by the use of addition agents, and their use is mainly confined to continuous plating of sheet, strip and wire where compensation can readily be made for the limited throwing power and where their high deposition rates are of particular advantage. The solutions are operated at room temperature with 100 per cent cathode efficiency.

The range of coating thicknesses deposited is 12–50 μm, at the lower end of which coating porosity can be a major adverse factor in corrosion-resistant performance. Coating porosity can be reduced and brightness improved by flow brightening after plating as with hot-dipped tin coatings (see page 74).

Electrodeposited tin is used as a protective coating on steel and copper alloy substrates but in the case of the latter materials the tin is cathodic and localised substrate corrosion will occur at discontinuities. Tin coatings are also used in electrical and electronic applications, particularly where good solderability is required.

Alloys of tin that are electrodeposited on steel for corrosion-resistant coatings are the 65/35 tin–nickel alloy and the 80/20 and 75/25 tin–zinc alloys. Tin–nickel deposits have a high hardness and good tarnish resistance; they are semi-bright and have a pinkish coloration. The alloy is deposited from an acid chloride/fluoride bath at 65–70°C using either alloy anodes or mixed anodes of both tin and nickel; copper undercoats are normally used to improve adhesion to steel substrates. Tin–zinc alloys have excellent solderability; they are plated from alkaline baths containing sodium stannate and either zinc cyanide or zinc carbonate, operated at 65–70°C with alloy anodes; the baths have good throwing power.

Zinc

Three classes of plating solutions may be used for electrodepositing zinc; they are acid, neutral and alkaline in character and in all cases pure zinc anodes are used. Deposit levelling properties are very poor.

Acid baths are generally based on zinc sulphate, although the chloride or fluoborate salts may be used and the brightness of the deposits can be enhanced by the use of addition agents in the plating bath. The acid baths are operated at about 30°C, cathode efficiency is around 100 per cent, but the throwing power is poor. Deposits from these baths tend to be softer and more ductile than those obtained from the alkaline baths.

Greatly improved throwing power is obtained by using the neutral chloride or pyrophosphate plating baths, though there is some loss of cathode efficiency, which ranges from 80 to 95 per cent. High rates of deposition can be obtained from the chloride bath, and fine-grained, pore-free deposits are produced by the pyrophosphate bath; bright deposits are obtained from both types of baths by the use of addition agents.

The alkaline baths are based on zinc cyanide; the properties of the baths and of the deposits obtained from them can be varied by altering the alkalinity of the bath and the free cyanide content. Very good throwing power and brightness of deposits are general characteristics of the cyanide baths, but cathode efficiency is reduced (falling in the range 75–95 per cent) and rates of deposition are lower than with other types of zinc plating bath.

Zinc is a widely used electrodeposition metal, being used for the protection of ferrous components ranging from small fasteners and fittings (which are barrel plated) through larger components for engineering applications (jig plated) to continuously plated sheet, strip and wire. Coating thicknesses may range from a few micrometres applied mainly for decorative purposes, and having only a limited degree of corrosion protection, to coatings of approximately 25 μm thickness, which give a long period of corrosion protection to the substrate by sacrificial corrosion. Thicker deposits can, of course, be produced by hot-dip galvanising or metal spraying.

Plating on plastics materials

Before leaving the subject of electrodeposited metals used as coatings it is necessary to consider the question of plating on plastics materials. Because of their special properties plastics materials can offer advantages over the use of metals as basic constructional materials through aspects of cost, lightness and general inertness and insulating properties. However, there are disadvantages to their use such as their lower mechanical strength and ductility, their degradation by the action of heat and light and their very different appearance compared with metals. For these reasons the application of metal coatings to plastics materials offers a means of producing a composite material that can have many uses in fields of consumer goods and light engineering.

Two of the basic properties of plastics materials complicate the application of metal coatings to them: because they are electrical non-conductors they cannot be directly electroplated, and because they are hydrophobic (water-repellent) treatment in aqueous chemical solutions is very difficult.

The first step in the coating of plastics materials with metals consists of treating them with a strongly oxidising acidic solution (such as a chromic/sulphuric/phosphoric acid mixture), which converts them to the hydrophilic (water-receptive) condition. This treatment also selectively etches the surface of the plastics material to produce micro-roughening, which provides a mechanical key to improve the adhesion of the subsequently deposited metal layers. However, with some types of plastics materials that do not have in their structure separate phases that will respond selectively to etching, it may also be necessary to apply an additional organic solvent treatment at this stage of processing.

After etching, the surface of the plastics material is conditioned by nucleating with discrete particles of metallic palladium by immersion in solutions of stannous chloride and palladium chloride, after which a continuous deposit of either copper or nickel can be obtained on the nucleated surface by immersion in an electroless plating solution (see page 82).

The electroless copper or nickel deposits, 1–2 μm thick, bond to the surface of the plastics material mainly by mechanical keying, though there is evidence to support theories of some chemical bonding between the metal and the plastics material. The electroless deposits provide the means of conducting electricity through the surface of the article so that further treatment in conventional electroplating processes can be carried out.

The two plastics materials most widely used for preparing plated plastics components are ABS (acrylonitrile-butadiene-styrene polymer) and polypropylene, although processes exist for treating a range of other plastics materials. Electroplated coatings of copper, nickel, chromium and composites of these metals are used (in combinations and thicknesses similar to those used on metal substrates) for decorative and protective purposes in the automotive, consumer hardware and electronic fields.

Because of the fact that when plastics materials are used as substrates for metal-coated articles their inertness and electrical non-conductivity stop them contributing to any corrosion reaction affecting the coating metals, deterioration of appearance of coated articles may well be less than with similarly plated metal articles, where corrosion products of the substrate metal are produced. For this reason it may well be entirely practicable to obtain satisfactory performance in service with thinner metal coatings than would be necessary for similarly plated metal articles. On the other hand, there is considerable difference in the rate of thermal expansion of metals and plastics; in applications involving temperature fluctuations, this imposes severe stresses on the metal/plastics composite; furthermore, the bond strength between the metal coating and the plastics substrate is low, and thermally induced stresses can lead to loss of coating adhesion unless special precautions are taken during processing. Specifically, it is often necessary to apply a first undercoat of a soft, ductile metal such as copper to a minimum thickness of 20 μm before applying the corrosion-protective coating, so that the ductile undercoat can accommodate thermal stresses and retain satisfactory adhesion. The lower mechanical strength of plastics materials compared with metals requires some design changes in compensation; in addition, it must be remembered that plastics have a low impact strength and a high notch-sensitivity. Because of this, premature failure can occur in sharply recessed areas if cracks are produced in the metal coatings applied to plastics (for example, chromium topcoats) through the stress-raising notch effect. For this reason it may well be necessary to amend design parameters still further if plastics components are required to be metal-coated.

Vapour deposition

The deposition of metals from the vapour phase is a method of producing coatings having properties that differ from those of coatings produced by other means. Thus it is possible to obtain coatings that have an extremely high degree of purity and freedom from oxides, are extremely thin, bright and non-crystalline, and can be deposited directly on to either metallic or non-metallic substrates. Coatings can be produced employing metals that cannot be deposited by other means, either because they cannot be electrodeposited from solutions or because they cannot be applied from the molten state (owing to excessively high melting points or excessive rates of oxidation during melting).

The basis of all vacuum deposition processes is treatment in an evacuated chamber containing the coating metal, which is vaporised, and the article to be coated. The degree of vacuum required for the successful operation of the process is moderately high, pressures of the order of 10^{-2} to 10^{-3} N/m^2 being needed. When the coating metal enclosed in the vacuum chamber is heated it passes into the vapour phase at a temperature considerably lower than its normal boiling point, and the vapour that fills the chamber condenses to form an even, solid coating on all cooler surfaces — the work to be coated and also the chamber walls.

The process is thus relatively simple and, since it is operated 'dry', there is no necessity for subsequent cleaning and drying operations. The articles being coated are subjected to only a very small increase in temperature during coating and the deposits are free from pores and inclusions. Some wastage of the coating metal occurs, because of condensation on the walls of the vacuum chamber as well as on the work to be coated, and the necessity to provide an adequate degree of vacuum imposes considerable capital investment in costly equipment for processing. The need to contain the articles to be coated wholly within an evacuated chamber also imposes some limitations on the size and number of them that can be processed in a single batch, but processes exist for vacuum coating as a continuous operation.

Two fundamentally different processes may be employed for producing vapour coatings. In the first of these the coating metal, in the form of bar or wire or contained in a crucible, is electrically heated to vaporisation by either resistance or arc methods. The metallic vapour molecules traverse the vacuum chamber in straight-line paths from their source and condensation occurs on any cool surface encountered in traversing these paths. Because of this mode of traverse it is necessary to rotate the work to be coated (so that all areas of a complex shaped article are presented to the vapour molecules) and/or to employ multiple vapour sources located in different parts of the vacuum chamber.

In the second method of application the heated coating source is given a high-voltage anodic charge and the work to be coated is cathodically charged. When this is done the anodically charged vapour molecules are attracted to the cathodically charged workpieces, discharge occurring on deposition. The method is known as cathode sputtering; it provides even deposits without the need for rotation of the workpieces within the chamber, and avoids wastage of the coating metal since condensation does not take place on the walls of the vacuum chamber. Auxiliary anodes may be employed within the chamber in order to accelerate the coating process and to exercise control over coating thickness in localised, recessed areas of the workpieces.

Normal good-quality cleaning and degreasing pretreatments are required for articles that are to be vacuum coated, concluding with very thorough drying before loading into the vacuum chamber. During pumping down of the vacuum chamber any gases entrapped in the workpieces will be drawn off. In order to minimise this outgassing and enable the working vacuum to be rapidly and readily attained, articles to be coated are sometimes sealed by lacquering prior to loading in the chamber; the coating metal is then deposited on the lacquered surface. When simple vacuum metallising is used both metallic and non-metallic articles are treated identically, but when cathode sputtering processes are employed it is necessary to pretreat non-metallic articles with conducting lacquers so that they will accept the required high-voltage electric charge.

Coating thicknesses deposited by vacuum metallising techniques can range from less than a nanometre up to tens of micrometres, and any metal capable of being vaporised in a vacuum can be applied. Aluminium coatings that are pore-free and of very high lustre are commonly used for both decorative and protective purposes, and protective coatings of zinc, cadmium, titanium and zirconium are also frequently produced. Precious-metal deposits are applied for highly decorative applications and in specialised applications in electrical, electronic and aerospace fields where very thin and highly protective coatings are an essential requirement. One special advantage of vacuum deposition is that coatings are produced without the generation of hydrogen, which can embrittle the substrate material; for this reason this method of coating finds applications in the protective coating of high-strength steels with zinc or cadmium when such coated materials are to be subsequently used in highly stressed conditions.

Diffusion coating

Diffusion coating processes enable the surface of a metal article to be changed in chemical composition by diffusing into it another metal or a non-metallic element. Alloying between the substrate and the coating metals occurs at the diffused surface, but little or no change of the dimensions of the diffused article takes place. Diffusion coatings can be applied to a range of metals and alloys including copper, molybdenum, nickel, niobium, tantalum, titanium and tungsten, but the widest use of these coatings is on ferrous materials.

Metals used as diffusion coatings for protecting steel against atmospheric corrosion and/or high-temperature oxidation are aluminium, chromium, silicon and zinc. Chromium and boron diffusion coatings are also used to improve wear resistance.

Metals that are to be diffusion coated are first cleaned by descaling and/or pickling and degreasing in a similar manner to that employed for preparing them for hot-dipping (see pages 67–68). They are then completely dried and heated either (a) in contact with the powdered coating metal in an inert atmosphere (solid-state diffusion) or (b) in an atmosphere containing volatilised compounds of the coating metal (gas-phase diffusion). The diffusion of the coating metal in the basis metal is on an atomic scale, and atomic and molecular bonding occurs so that total adhesion is obtained and the distortion of the crystal lattice that occurs leads to an increase in the hardness of the metal surface. Coating thicknesses applied are generally in the range 5–15 μm.

Solid-state diffusion

Coating treatment is carried out in a sealed container, with the cleaned metal articles packed in powder containing the coating metal; the container is heated for several hours at a temperature near to but lower than the melting point of the coating metal. Zinc coatings on steel are known as Sheradised coatings, and the diffusion layer is an alloy containing 8–9 per cent of iron in the zinc. Aluminium coatings on steel or copper are known as Calorised coatings; in these, aluminium oxide forms in all the surface layers that contain more than 8 per cent of aluminium; this oxide film imparts high corrosion resistance but also severely embrittles the surface layers, and post-calorising annealing treatments are given.

Gas-phase diffusion

Chromium and silicon are most commonly used for diffusion coating from the gaseous state. The vaporised halide of the coating metal is passed over the metal to be coated while it is being heated and maintained in either an inert or a reducing atmosphere. Three basic reactions occur:

(a) an exchange reaction between the two metals;
(b) a reduction of the coating halide to produce the metallic state;
(c) dissociation of the coating halide to produce the metallic phase.

These reactions are encouraged to proceed to completion by maintaining the concentration of available halide and by removing the reaction products from the reaction vessel.

As chromium is diffused into steel the microstructure is converted to the ferritic form. The coatings generally have a columnar microstructure, and exhibit corrosion- and oxidation-resistant properties allied with improved wear resistance. With silicon diffusion coatings oxidation- and acid-resistance are obtained, and the surface is very hard and brittle.

Mechanical application

Coatings may be applied mechanically by roll or extrusion bonding or by forging. In all three cases the mechanism is the same, namely pressure welding between the coating and the substrate. The major problem in achieving good adhesion between the coating and the substrate metals is the complete elimination of contaminants (in the form of oxides) from the interface by applying adequate pressure in such a manner as to break down any oxide particles and disseminate them in the plastically deformed metals.

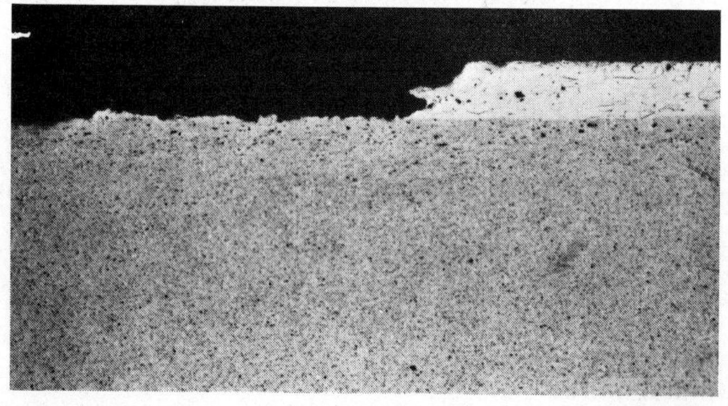

Figure 3.10 Sacrificial corrosion of aluminium–1 per cent zinc cladding on aluminium–1¼ per cent manganese alloy (× 50)

Very thin coatings of soft, noble metals such as gold can be produced by hammer-beating on to a harder substrate, but this method of application is restricted in practice to high-cost, craftsman-produced decorative finishes. At the more practical and more commonly used level, roll-bonded and extrusion-clad materials are produced for a variety of applications.

Sheet and simple-shaped extrusions of aluminium alloys having a lesser corrosion resistance than pure aluminium may be produced clad with high-purity aluminium or with another aluminium alloy having greater corrosion resistance than the core alloy. When this is done the composite material will have a long service life in a corrosive environment since the coating can provide sacrificial protection to the core metal, as shown in *Figure 3.10*.

Composite extruded tubes may be produced for use in heat-exchangers, with combinations of inner and outer components consisting of copper alloys, nickel alloys, mild steel or stainless steel. Copper or aluminium cables may be given extruded sheaths of lead or lead alloys or sheaths of pure aluminium. Finally, steel sheet can be clad with roll-bonded lead, which offers high resistance to atmospheric or acidic corrosive environments and which also has the property of a high sound-damping capacity.

The range of coating/substrate thickness combinations that can be produced by cladding or extruding is extremely wide and can be very accurately controlled; the coatings produced in this way are completely free from pores or other coating discontinuities.

4

Coating performance

The corrosion protection given by a metal coating depends upon a number of factors. Each of these factors must be considered individually, but they must also be considered in relation to each other since the overall effects vary with the possible combinations of the individual factors. The parameters that must be considered are as follows:

(a) the specific corrosive environment encountered
(b) the substrate on which the coating is applied
(c) the coating metal employed
(d) the type of coating, as governed by its method of application
(e) the presence of any under- or over-coatings

The specific environment and the substrate material are usually the fixed conditions in any given application and in order to evaluate coating performance, and hence the selection of the most suitable coating system, it is probably best to consider the performance of individual coating metals in relation to these two factors, using the natural environmental variables already given in *Figure 1.2*.

Aluminium

In its pure form aluminium offers a high degree of corrosion resistance to the atmosphere owing to the fact that a thin, tenacious oxide film is rapidly formed on exposure to air. This oxide film is inert and its formation rapidly stifles further corrosion over the metal surface. In an industrial environment the corrosion rate of aluminium, averaged over a six-year period, is 2–5 μm per year but the rate of penetration in the sixth year is only one quarter that which occurs in the first year. By comparison, mild steel corrodes at a rate of 20–25 μm per year and this rate of penetration is substantially constant irrespective of the length of exposure.

Such corrosion as does occur with aluminium takes the form not of general surface wastage, as with steel, but of localised pitting initiated at weak points in the oxide film. Typically, these small pits may reach a depth of 0.25–0.5 mm after six to seven years' exposure to an industrial environment but only about one tenth of that depth in a rural or marine environment. Pitting of this nature has little effect on overall strength but can adversely affect the performance of thin sheet or thin coatings as a result of perforation.

Although a long service life can thus be achieved with pure aluminium coatings (provided that they are sufficiently thick to prevent the corrosion pits penetrating completely and so exposing the substrate metal), the slight surface roughening and localised pitting causes increased retention of dirt and general deterioration in aesthetic appearance, which might make them unacceptable in applications involving decorative as well as protective considerations.

If aluminium coatings are applied by metal spraying, as opposed to cladding, the increased oxide content of the metal tends to reduce the number of active corrosion sites and the rate of corrosion, hence increasing the effective service life of the coating. Porosity, which is also greater in sprayed coatings, may decrease life in that it allows easier access of the corrosive environment, but this may be offset by the ability of the pores to retain coating corrosion products which can stifle further attack.

Aluminium coatings applied by hot-dipping also have heavier oxide films on their surface than clad coatings and hence have greater intrinsic corrosion resistance. Properly applied they are free from porosity. The alloy layer formed between the pure aluminium topcoat and the steel substrate ensures adhesion and prevents any risk of corrosion spreading along the coating/substrate interface if the substrate is exposed at local pitting corrosion sites; attack spreading along the interface can sometimes occur, however, with clad or sprayed coatings.

Under immersed conditions the rate of corrosion of aluminium is dependent upon the dissolved oxygen content of the water, on its chloride content and, in particular, on the presence of heavy metals such as copper. The nature and quantity of scale-forming salts in the water also affect the corrosion rate. Very high chloride contents cause rapid general corrosion, and hence aluminium is generally unsuitable for applications involving immersion in sea water. In potable waters deep pinhole pitting can readily occur in the presence of very small quantities of copper dissolved in the water, and hardness scales deposited over the pits enable the micro-environment within the pit to remain active so that the corrosion rate does not materially decrease with increasing time of exposure. If, however, the temperature of the water is increased to approximately 80°C pitting corrosion is prevented — probably as a result of the precipitation of heavy metals and hardness salts and a reduction in the amount of dissolved oxygen.

Aluminium coatings may be seen, therefore, to be suitable for use in soft, pure waters at any temperature and in other waters if used hot. Its use in cold, hard waters gives only limited life.

In soils or chemical environments aluminium coatings perform satisfactorily in conditions where sufficient oxygen is available to allow ready formation of the protective alumina film and where conditions are slightly acidic, but they are readily attacked in alkaline environments. The highly protective nature of the air-formed alumina film also ensures excellent resistance to the high temperatures that are encountered in exhaust flues, even in the presence of acidic products of combustion.

If and when penetration to the substrate occurs the performance of an aluminium coating in any environment depends on the nature of the substrate. Aluminium is not normally employed as a coating metal in combination with additional coating layers, and the only two substrate metals commonly coated with aluminium are steel and other aluminium alloys. Aluminium in contact with steel may be either weakly anodic or cathodic according to the environmental conditions, so sacrificial protection or enhanced corrosion of any exposed steel will be only minimal; continuing attack may well be governed by the ease with which ferrous corrosion products are removed from the corrosion pits.

Many aluminium alloys (notably those containing copper, zinc and magnesium) are less resistant to corrosion than pure aluminium but are also susceptible to specialised forms of attack such as stress-corrosion cracking and intercrystalline corrosion. However, since these alloys are often cathodic (more positive) to pure aluminium they can be protected by sacrificial action if coated with the pure metal; the composite also exhibits the greater intrinsic corrosion resistance of the pure coating while retaining the greater mechanical strength of the alloy core metal. Both clad and sprayed coatings of this type have been used to ensure long service lives for aluminium alloy components exposed to the atmosphere or immersed in potable waters.

Because aluminium relies on the ready formation of a protective oxide film in order to achieve high corrosion resistance it follows that restricted availability of oxygen will impair that corrosion resistance. For this reason aluminium is very susceptible to enhanced corrosion in crevices or other regions where moisture can be entrapped and oxygen supply limited. Care must always be taken to avoid crevice conditions in practical applications and to exclude moisture, by using sealants during assembly, from any crevices that cannot be designed out. A further hazard is the greatly accelerated corrosion induced on aluminium by contact with dissimilar metals. Copper and its alloys are the most aggressive in this respect, but steel can also act in this way particularly if large areas of bare steel are coupled to small areas of aluminium. Ideally, therefore, such bi-metallic contacts should be completely avoided; where this is not possible the dangers of attack on the aluminium can be reduced by the use of nickel or cadmium coatings on the steel and sealants to prevent ready access of the corrosive environment to the joint. Apart from direct bi-metallic contact with copper and its alloys it is often not realised that enhanced corrosion of aluminium can readily occur if the run-off of corrosion products from the copper comes into contact with the aluminium, e.g. rainwater passing from copper roofing on to aluminium window frames.

Typical applications of aluminium coatings are summarised in *Table 4.1*.

Table 4.1 TYPICAL APPLICATIONS OF ALUMINIUM COATINGS

Substrates	*Applications*	*Coating methods*
Steel	Structure exposed to the atmosphere, immersed in water or buried. Components to resist high-temperature oxidation or hot flue gases	Hot-dipping, or metal spraying, or cladding
	Decorative finishes	Vacuum deposition
Aluminium alloys	Structures or components exposed to aggressive atmospheres, immersed in water or buried (applicable particularly to the protection of aluminium alloys from stress-corrosion)	Metal spraying or cladding
Plastics	Decorative finishes (particularly for reflective finishes)	Vacuum deposition

Cadmium

The rate of corrosion of cadmium when exposed to a corrosive environment is generally linear with time, although this may be modified by the nature of the corrosion products produced in different types of environment. Cadmium provides sacrificial protection when used as a coating on steel, the coating life being proportional to its thickness (see page 38).

In exposure to a severe industrial atmosphere a 25 μm thick cadmium coating protects steel for a period approaching one year, but in a marine environment the life may be extended to about five years. The reason for this difference is that cadmium sulphates produced by corrosion in a polluted industrial atmosphere are soluble and are removed by rain, whereas in a marine atmosphere insoluble carbonates and basic chlorides are produced, which tend to be retained on the surface thus reducing the rate of subsequent corrosion.

Cadmium also provides good corrosion protection to steel in conditions where condensation can occur in enclosed spaces (particularly where organic vapours may be present), in immersion in stagnant or soft neutral waters, and in applications involving exposure to alkaline or acidic environments. In all these applications its use as a coating for steel is preferred to that of zinc.

Cadmium coatings are more tarnish-resistant than zinc and so retain a clean, attractive appearance for longer periods. Solderability of cadmium is good, but the toxicity of its vapour precludes its use for coated components that have to be subsequently welded, and the toxicity of cadmium metal and of its salts prevents its use in contact with foodstuffs.

Mention has already been made, under 'Aluminium', of the benefits of cadmium coatings for steel that must be assembled in contact with aluminium; cadmium also has a low torque resistance, which is beneficial for coated-steel threaded components that have to be regularly assembled and dismantled.

Typical applications of cadmium coatings are summarised in *Table 4.2*.

Table 4.2 TYPICAL APPLICATIONS OF CADMIUM COATINGS

Substrates	*Applications*	*Coating methods*
Steel	Structures and fasteners exposed to humid atmospheres or to organic vapours. Surfaces requiring good solderability. Low-torque threaded fasteners. Components in bi-metallic contact witn aluminium	Electrodeposition or vacuum deposition

Chromium

The commonest use of chromium as a coating material is in the form of electrodeposits, which remain virtually inert on exposure to the atmosphere or when immersed in waters. Because of its high degree of resistance to corrosion and tarnishing, together with its colour and high lustre, it is principally used for decorative finishing, although its hardness of 800–900 HV makes it an eminently suitable material for wear-resistant coatings.

Thin, decorative deposits of chromium are always porous; because of internal stress and brittleness in electrodeposits, porosity cannot be eliminated by increasing the deposit thickness since spontaneous cracking occurs. The discontinuities that occur enable corrodents to penetrate the coating and attack the underlying metals, the chromium surface providing a large cathodic area so that rapid localised corrosion occurs on the exposed underlying (anodic) metals. For this reason, chromium is almost always used in conjunction with suitable corrosion-resistant undercoatings such as nickel rather than as a single coating material. The exceptions to this are articles requiring a cheap, decorative finish that will be subjected only to the mildest of corrosive conditions (such as fittings for indoor use), and components plated with hard chromium for wear-resistant applications. Although thick deposits of hard chromium always contain cracks, the electrolyte path to the substrate is narrow, tortuous and often not fully continuous; even so, protective undercoats may still be necessary in applications where the more corrosive environments are encountered, such as hydraulic equipment exposed to immersion in mine waters.

Ordinary decorative chromium electrodeposits are usually approximately 0.3 μm thick. When these are used with nickel undercoats of suitable thickness and quality, substrates of steel, zinc alloy or copper can be completely protected for periods of outdoor atmospheric exposure of about six weeks to six months. After this period of time small pits or blisters containing corrosion products of the substrate metal occur and the decorative appearance of the plated article becomes unacceptable, although its functional use may remain unimpaired for greatly extended periods. Some, minimal, improvement in these acceptable lives can be achieved with the thicker ‘crack-free’ deposits (see Chapter 3) but excessive brittleness is an attendant danger with these deposits. If, however, micro-discontinuous chromium deposits such as micro-cracked or micro-porous chromium are employed (see Chaper 3) in thicknesses in the range 0.3–1.0 μm, according to the process of electrodeposition used, the lower local anode current density delays penetration of corrosion through the protective nickel undercoats to such an extent that the fully acceptable decorative life may range from one to five years. Even after these periods of time, loss of appearance is frequently not occasioned by corrosion of the substrate metal, but rather by dulling of the chromium surface as minute micro-flakes of chromium become detached from the nickel as a result of superficial corrosion of the nickel.

Hard chromium deposits for wear-resistant applications may range in thickness from as little as 10 μm to as much as 400 μm; beyond this upper limit excessive brittleness may cause spalling of the coating in use. The cracks present in these hard deposits can fulfil a useful function in service, in that they tend to retain lubricants applied to the component to reduce still further friction and wear.

Chromium coatings are unsuitable for protection against strongly acidic environments, acids such as hydrochloric attacking the metal vigorously and stripping the coating from the substrate.

Chromium used as a diffusion coating material for steel is primarily intended for increasing the oxidation resistance and hardness and wear resistance of the coated article; the fact that the corrosion resistance to aqueous environments is somewhat increased by the coating process is of secondary importance. The beneficial effects of the chromium are obtained by its alloying action with the steel, producing iron–chromium solid solutions and intermetallics and chromium carbides, and producing a ferritic microstructure in the metal. Hardness values up to 1800 HV can be obtained in this way. The effect of chromising on corrosion resistance depends upon the composition and depth of the diffused layer. Greatly improved corro-

Table 4.3 TYPICAL APPLICATIONS OF COPPER COATINGS

Substrates	*Applications*	*Coating methods*
Steel	Decorative overlay to protective nickel coatings on components exposed to the atmosphere. Wear-resistant coatings on engineering components, e.g. rollers, hydraulic rams, printing cylinders	Electrodeposition
	Hard, wear-resistant coatings on engineering components. Coatings resistant to high-temperature oxidation	Diffusion coating
Aluminium, copper and its alloys, zinc alloys	Decorative overlay applied either directly or over protective nickel undercoats on components exposed to the atmosphere	Electrodeposition

sion resistance can be obtained if a solid solution of iron–chromium is produced, but preferential attack can occur on this solid solution if a precipitate of chromium carbide particles is formed. The presence of chromium diffusion coatings improves the oxidation resistance of steel, and protection can be retained for long periods of exposure to temperatures up to 750°C.

Typical applications of chromium coatings are summarised in *Table 4.3.*

Copper

Copper and copper alloys offer a very high degree of corrosion resistance when exposed to the atmosphere, owing to the formation of a dark surface film which is principally composed of cuprous oxide together with basic salts derived from the other alloy constituents. The attack is uniformly distributed over the surface area, and the rate of penetration ranges from 0.2–0.6 μm per year in a rural environment to 0.9–2.2 μm per year in an industrial environment. After a period of some six or seven years' exposure to marine or industrial environments, many copper alloys develop a green patina owing to the formation of basic copper chlorides or sulphates. Patination of this type is a familiar phenomenon that is decoratively acceptable, and when fully developed it is a very stable condition leading to extremely long service life.

Copper and its alloys also have high ductility and good electrical and thermal conductivities, properties that greatly enhance their value as coating materials. When copper is used as an electrodeposited coating, the high degree of levelling obtained can be exploited to reduce the amount of pre-plate polishing that would otherwise be required to produce highly decorative finishes.

Although the rate of corrosion of copper in the atmosphere is slow, its initial bright appearance is rapidly lost by tarnishing. The naturally produced copper oxide films that are responsible for the slow corrosion rate of the metal are darker and duller than the film-free metal, and if traces of sulphides are present in the environment complete blackening rapidly occurs. For these reasons decorative copper finishes must be protected by applying clear lacquer finishes (frequently containing inhibitors such as benzotriazole) to exclude the environment and so prevent tarnishing and retain a pleasing, bright appearance.

Under conditions of immersion in sea water or potable waters, films of basic copper salts are produced on the surface of the metal; the rate of corrosion is then reduced to a very low level, provided that the rate of flow of water over the metal surface is not sufficiently high or sufficiently turbulent to break down the films and remove them from any portion of the surface. If this should occur the areas of film-free metal so produced are anodic to the (larger) surrounding areas of filmed metal and intense localised corrosion occurs; such corrosion is known as impingement attack (see *Figure 4.1*).

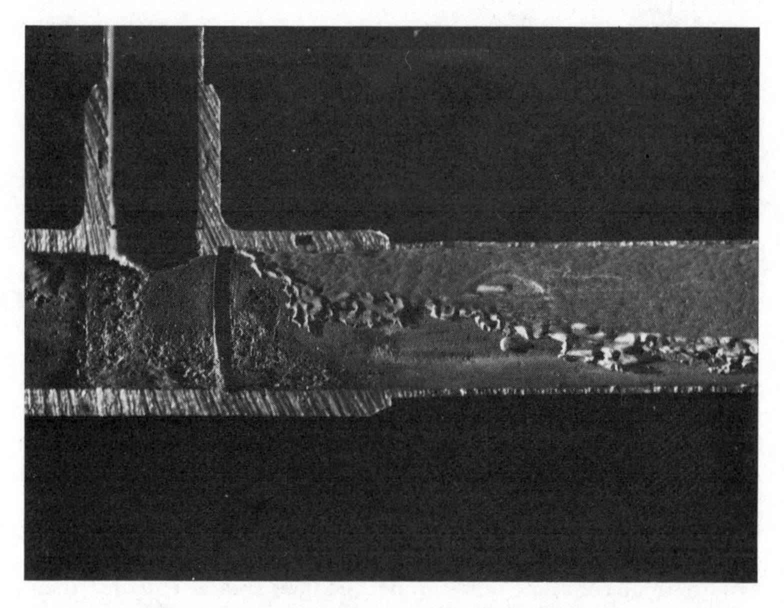

***Figure 4.1** Impingement corrosion in copper water-pipe (× ½)*

Protective films of basic copper salts also provide good protection to copper buried in soils, though this is impaired if acidic conditions occur. A further hazard, both in soils and waters, is the presence of small quantities of ammonia, which greatly accelerates attack on the metal and, in the presence of stress, produces rapid stress-corrosion failures.

With copper alloys corrosion may take the form of selective removal of the alloying metal leaving a weak, porous matrix of impure, spongy copper. This form of attack is known as *dezincification* in brasses (removal of zinc), *dealuminification* in aluminium bronzes (removal of aluminium) and so on according to the alloy, and can occur in both atmospheric exposure and under immersed conditions. The attack is generally stimulated by a deficiency of oxygen in the corrosive environment, so it frequently occurs in creviced areas of a component or under deposited silt.

If copper is used as a coating metal on steel it is cathodic to the steel substrate and so stimulates attack at any coating discontinuities. For this reason it is most widely used as an undercoat material in conjunction with other, more noble materials.

Typical applications of copper coatings are summarised in *Table 4.4*.

Table 4.4 TYPICAL APPLICATIONS OF COPPER COATINGS

Substrates	*Applications*	*Coating methods*
Steel	Decorative and protective coatings resistant to atmospheres or water immersion. Surfaces requiring good solderability. Surfaces requiring good electrical conductivity	Electrodeposition, or electroless plating, or cladding
Steel or zinc alloys, aluminium	Undercoat for protective nickel/chromium coatings. Engineering coatings for printing, engraving or electronic applications	Electrodeposition or electroless plating
Plastics	Preliminary coatings for protective plating of plastics	Electroless plating and electrodeposition

Gold

Gold is the noblest of the metals and is completely resistant to corrosion and tarnishing by all environments except aqua regia. For this reason it would provide the best coating material for complete corrosion control except, of course, for its high cost. Because of the cost factor coatings of gold, where they are used, are kept to a minimum thickness and porosity can be a major hazard. Where porosity does occur, the highly cathodic nature of the gold topcoat stimulates localised attack on any substrate material exposed at the discontinuities. Consequently it is necessary to make a careful choice of substrate and/or undercoat compatible with the environment, so that such attack as does occur does not adversely affect the appearance or performance of the coated component.

Gold has very good electrical conductivity, and because of its high corrosion resistance and tarnish resistance it retains a low electrical contact resistance for indefinite periods. The pure metal is soft and ductile but the hardness, and hence the wear resistance, can be improved by alloying additions. Gold and gold alloys in the form of thin electrodeposits are widely used for decorative finishes.

Typical applications of gold coatings are summarised in *Table 4.5*.

Table 4.5 TYPICAL APPLICATIONS OF GOLD COATINGS

Substrates	*Applications*	*Coating methods*
Copper and its alloys	Decorative and protective coatings for jewellery. Protective coatings for aerospace hardware. Protective coatings with good electrical conductivity for electronic applications	Electrodeposition, or electroless plating, or cladding, or vacuum deposition
Plastics	Electrically conducting coatings	Electroless plating or vacuum deposition

Lead

Lead offers a very high degree of corrosion resistance to industrial atmospheres and to soils and waters. Its protective action is due to the ready formation of insoluble sulphates, carbonates, sulphides and oxides, which adhere to the metal surface and so stifle further corrosion. Lead is particularly inert in acidic environments, but its corrosion resistance is impaired when chlorides are present owing to the greater solubility of lead chloride.

The extreme softness and high ductility of lead are extremely beneficial when the metal is used as a coating material for steel. If the composite is subjected to mechanical damage the soft lead coating tends to smear over the damaged surfaces, so the substrate is not readily exposed to the corrosive environment. Protection is thus retained by continued exclusion of the environment; the lead is incapable of providing sacrificial protection to the steel substrate. A further property of lead that can be of benefit when it is used as a coating material (though not in a corrosion-control context) is its high sound-damping capacity; lead-coated composites are very useful for sound-insulation applications.

Typical applications of lead coatings are summarised in *Table 4.6.*

Table 4.6 TYPICAL APPLICATIONS OF LEAD COATINGS

Substrates	*Applications*	*Coating methods*
Steel or copper	Acid-resistant coatings for chemical plant. Structures resistant to atmospheres, waters or buried conditions. Surfaces requiring good solderability. Sound-damping applications	Hot-dipping, or cladding, or electrode position

Nickel

Nickel is a white metal comparable in hardness with steel and with a high degree of corrosion resistance to the atmosphere and to waters. The corrosion rate in atmospheric exposure is of the order of 0.02–0.2 μm per year and tends to lessen slightly with increasing period of exposure owing to the passivation of the metal surface through the formation of inert oxide films. Although nickel is a ductile metal the ductility of nickel coatings is very dependent upon their method of production and their purity; many electrodeposited nickel coatings (particularly the organic bright nickels) may be brittle and have a high degree of internal stress. Similarly, chemically deposited nickel coatings have greater hardness, brittleness and different corrosion characteristics because of the incorporation of phosphorus or boron in the deposits — inherent in the method of deposition from complex solutions.

When nickel is freely exposed to a corrosive environment tarnishing rapidly occurs and shallow, general corrosion develops widespread over the surface. Consequently, plain nickel coatings are very efficient for protecting steel in engineering applications where decorative appearance does not have to be maintained. Resistance to attack by acids is also particularly good. In decorative coating applications, however, the rapid tarnishing is undesirable, and protective nickel coatings are generally given a decorative bright chromium overlay to retain an acceptable appearance. When this composite coating system is used, localised pitting attack occurs on any nickel exposed at discontinuities in the chromium topcoat and rapid penetration of the nickel can occur.

A 25 μm nickel coating may be penetrated by corrosion at a discontinuity in a chromium coating in as little as six weeks exposure to an industrial atmosphere because of the high anodic current density induced on the nickel at the break in the large cathodic chromium surface.

The reactivity of nickel electrodeposits varies with their purity, the organic bright nickels containing sulphur being more readily corroded than the purer semi-bright or dull nickels. This property is utilised in practice by depositing duplex coatings, with dull or semi-bright nickel adjacent to the substrate metal and an upper layer of bright nickel beneath the chromium topcoat. When corrosion occurs in these duplex nickel layers the attack is preferential on the upper (bright) layer, and penetration through the lower (semi-bright) layer is retarded at the expense of some lateral spread of the corrosion pit in the upper layer as shown in *Figure 4.2.* By using a composite nickel coating of this nature the rate of penetration of a 25 μm thick coating

Figure 4.2 Lateral spread of corrosion in bright-nickel layer of duplex nickel electrodeposit (× 750)

may be more than halved when used below a decorative chromium overlay, or reduced to less than one fifth or better when used with a micro-discontinuous chromium overlay.

When nickel is used as a coating over steel, zinc alloy or copper, preferential attack occurs on the substrate metal when the nickel layer is penetrated, and the coating may be undermined or blistered away from the substrate as corrosion products either exude from the pit or are retained *in situ* as shown in *Figure 4.3.*

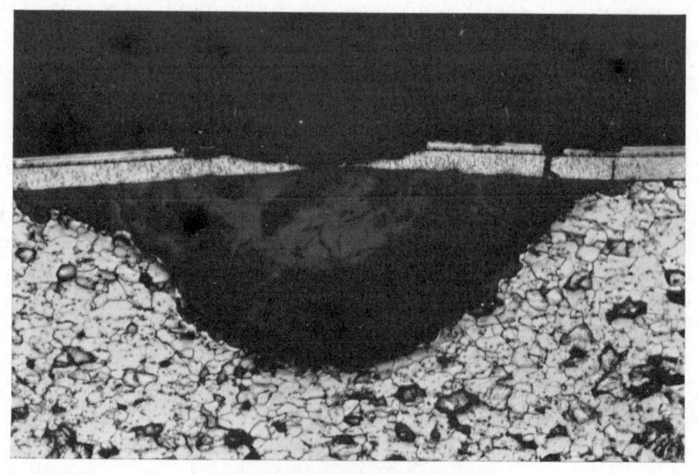

Figure 4.3 Preferential corrosion of steel substrate at pit penetrating through nickel + chromium electrodeposits (× 140)

Nickel also offers a high degree of resistance to elevated-temperature oxidation and to corrosion when immersed in sea water or potable waters, because of the ready formation of thin, tenacious oxide films that protect the metal from further attack. As with most of the metals that rely for their general corrosion resistance on passivity by a readily-formed oxide film, this property is impaired in situations where the corrosive environment is deficient in oxygen; consequently, pitting of nickel occurs in crevices or under conditions of exposure buried in soils.

Typical applications of nickel coatings are summarised in *Table 4.7.*

Table 4.7 TYPICAL APPLICATIONS OF NICKEL COATINGS

Substrates	*Applications*	*Coating methods*
Steel or zinc alloys, copper and its alloys, aluminium	Corrosion-resistant coatings for components or structures exposed to atmospheres or immersed in waters or sea water (used either alone or in combination with copper undercoats and/or chromium overlays). Protective coatings for chemical plant. Hard, wear-resistant coatings for engineering applications	Electrodeposition, or metal spraying, or cladding, or electroless plating
Plastics	Preliminary coatings for protective plating of plastics	Electroless plating and electrodeposition

Silver

Although silver is one of the noble metals and, as such, has a generally high corrosion resistance its use as a coating metal alone is not great since it is highly prone to severe tarnish blackening by minute traces of sulphur compounds in any corrosive environment. In order to retain a bright, decorative appearance silver coatings are frequently overplated with an extremely thin deposit of rhodium. Typical applications of silver coatings are summarised in *Table 4.8.*

Tin

Tin is very resistant to atmospheric corrosion, the rate of penetration ranging from 0.02 μm per year in a rural environment, through 0.1 μm per year in an industrial environment, to 0.25 μm per year in a

Table 4.8 TYPICAL APPLICATIONS OF SILVER COATINGS

Substrates	*Applications*	*Coating methods*
Copper and its alloys, nickel and its alloys	Decorative coatings for jewellery, cutlery and household articles. Protective coatings for chemical plant. Components for electronic applications	Electroless plating, or electrodeposition, or vacuum deposition
Plastics	Decorative coatings. Coatings for good electrical conductivity	Electroless plating or vacuum deposition

marine environment. The initial bright appearance is retained for long periods (although, in applications involving high humidity, deliquescent corrosion products retained on the surface detract somewhat from the appearance of the metal). Very little tarnishing is produced by sulphur contamination of the atmosphere.

In bi-metallic contacts tin is generally anodic to copper and iron, and cathodic to zinc and aluminium, although the exact bi-metallic relationship may vary somewhat with changes in the corrosive environment. Resistance to alkalis is poor owing to solution of the oxide film, but attack by acids is slow, particularly in the absence of a plentiful supply of oxygen. The resistance to acid attack is particularly good in the case of exposure to organic acids; this is of particular significance in the use of tin as a coating metal for steel for food cans, because a long service life is obtained, the steel substrate is anodically protected by sacrificial corrosion of the tin, and the tin itself is non-toxic.

Tin is virtually unattacked by immersion in potable waters and only very slowly attacked by sea water (average penetration 0.07–0.2 μm per year).

Tin is soft and extremely ductile; consequently tin-coated metals can be heavily worked without producing discontinuities in the

Table 4.9 TYPICAL APPLICATIONS OF TIN COATINGS

Substrates	*Applications*	*Coating methods*
Steel, copper and its alloys	Protective coatings for resistance to atmospheres, immersion in water or organic acids. Food canning. Surfaces requiring good solderability and electrical conductivity	Hot-dipping or electrodeposition

coating. Indeed, in some applications flow induced in the tin coating by heavy mechanical working may tend to close up porosity present in the 'as deposited' condition. Tin has excellent solderability and electrical conductivity.

Typical applications of tin coatings are summarised in *Table 4.9.*

Zinc

Zinc is a soft metal of low strength that corrodes freely but relatively slowly in the atmosphere at a comparatively constant rate. The corrosion rate in an industrial atmosphere is about 15 μm per year, and falls to about one fifth of that value in a marine or rural environment. The reason for the low corrosion rate is the tendency to produce basic zinc chloride and carbonate corrosion products, which stifle the attack. In conditions of high humidity superficial corrosion produces white corrosion products of zinc, which adhere to and stain the surface of the metal (a phenomenon known as white rusting) but this defect can be avoided by passivating the zinc surface by chromate treatment.

Because the corrosion rate is relatively constant the life of a zinc coating is generally proportional to its thickness; it is also more or less independent of the method of application — hot-dipped, metal sprayed or electrodeposited zinc all having similar corrosion rates. However, some slight variation in corrosion rate can occur with different forms of zinc coating; the iron–zinc alloy layer in a hot-dipped coating tends to corrode more slowly than pure zinc, and the porous nature of a sprayed zinc coating may entrap corrosion products and so progressively stifle the corrosion reaction.

Zinc provides a very good anode when coupled to steel and so gives efficient sacrificial protection when used as a coating for steel. Protection can be maintained over quite large areas of exposed substrate. For example, no measurable attack can be found on a 12 mm diameter area of exposed steel in a zinc-coated sheet even after as long as seven years' atmospheric exposure. Similarly, good sacrificial protection is given by zinc coatings applied to aluminium alloys, the coatings being applied by metal spraying.

The good anodic reaction of zinc coupled to steel makes it an excellent coating material for applications involving immersion in sea water or potable waters or when buried in soils. The rate of consumption of the zinc is increased in sea water or other high chloride waters, but in scale-forming neutral waters basic salts are produced, which slow down the rate of corrosion. Danger can arise from contact with copper when attack on the zinc is greatly stimulated, and similar problems can occur if copper salts in solution are precipitated as metallic copper on the zinc surface. In soft waters or acidic environments the rate of consumption of zinc is excessive.

Typical applications of zinc coatings are summarised in *Table 4.10.*

Table 4.10 TYPICAL APPLICATIONS OF ZINC COATINGS

Substrates	*Applications*	*Coating methods*
Steel, aluminium alloys	Protective coatings for structures and fasteners or components exposed to atmospheres, immersed in water or buried	Hot-dipping, or electrodeposition, or metal spraying, or vacuum deposition

5

Selecting a coating

The selection of the best coating or combination of coatings for any particular corrosion control application necessitates consideration of all the factors involved so that the most economic choice may be made consistent with the performance required. The order in which these factors should be considered is likely to be as follows:

1. The environments(s) that will be met in service
2. The service life required
3. Decorative appeal and the degree of deterioration that can be tolerated
4. The substrate material
5. The shape and size of the article
6. Any subsequent fabrication
7. Mechanical factors

However, in any particular case the order of importance may be changed to meet special circumstances.

Environmental factors

Consideration of the corrosive environment or environments that will be met in service generally first results in the exclusion of unsuitable coating materials, leaving a number of materials of greater or lesser merit depending on the other requirements of the application. Thus aluminium would be ruled out of consideration as a coating metal in strongly alkaline environments, aluminium and lead would be unsuitable for use in high-chloride environments, copper and zinc would be unsuitable in acidic environments — in all cases owing to their excessive rates of corrosion in these environments. Aluminium, copper, nickel and tin are resistant to atmospheric environments; aluminium and nickel are also resistant to elevated temperatures but are attacked under conditions of limited access of oxygen. Nickel, copper and tin are resistant to potable waters and sea water but aluminium is less resistant to waters, particularly when the chloride content is high. Cadmium is preferable to zinc in humid environments containing organic vapours, and aluminium, nickel and tin offer good resistance to acidic environments. Lead gives good performance in atmospheric, acidic or aqueous environments, but not in the presence of high concentrations of chlorides. Chromium remains bright and unattacked in most environments, except acidic chloride environments, but coating discontinuities may allow preferential attack on undercoats or substrate metals; on the other hand, zinc and cadmium (which are electronegative to steel in the atmosphere and in waters) can provide efficient sacrificial protection to suitable substrates — notably steel. Silver, copper and, to a lesser extent, nickel are attacked by sulphur compounds, which produce unsightly and non-protective films on their surfaces.

Service-life requirements and acceptable deterioration

The service life required affects both the choice of coating metal and also its thickness, the latter being also dependent on the severity of the environment to which it is to be exposed. It would be uneconomic, for example, to apply a coating of a highly resistant metal to a component that is required to have only a very limited service life, unless it is essential to retain an initial decorative appearance throughout that life or in cases where any risk of failure through defects could not be tolerated for any reason — e.g. safety considerations.

The ways in which choice of coating metals and their thicknesses are affected by the service environment and service life may be illustrated by reference to *Tables 5.1* and *5.2.* Thus bright nickel is unacceptable for use outdoors in exceptionally severe corrosive conditions; reduction of 15–50 per cent in minimum nickel thickness requirement is allowed as the severity of the environment is reduced through four environmental categories; 12–16 per cent reduction in nickel thickness is also allowed if micro-discontinuous chromium is used instead of regular chromium in outdoor service. Similarly, somewhat greater thicknesses of aluminium on steel than of zinc on steel are recommended because of the poorer sacrificial action of aluminium; thickness requirements for both types of coatings must be increased with increasing life requirements or increases in the severity of the corrosive environment.

Table 5.1 TYPE AND THICKNESS REQUIREMENTS FOR NICKEL + CHROMIUM COATINGS ON STEEL

Service condition No.	*Description of typical environment*	*Coating type and thickness*	
		Ni	Cr
4	Exceptionally severe outdoor	40 μm duplex 30 μm duplex	0.3 μm regular 0.3 μm micro-discontinuous
3	Normal outdoor	40 μm bright 30 μm bright 30 μm duplex 25 μm duplex	0.3 μm regular 0.3 μm micro-discontinuous 0.3 μm regular 0.3 μm micro-discontinuous
2	Indoor with condensation	20 μm bright or duplex	0.3 μm regular or micro-discontinuous
1	Indoor dry	10 μm bright or duplex	0.3 μm regular or micro-discontinuous

Abstracted from BS 1224:1970 *Electroplated coatings of nickel and chromium*

Effect of substrate material

The particular substrate material that has to be protected by a metal coating influences both the choice of coatings and also possibly the methods by which they are to be applied. Zinc and cadmium are highly effective coating metals for steel, since they are anodic to steel and provide sacrificial protection to the substrate at discontinuities in the coating. Coatings that are cathodic to a substrate metal must be applied and maintained free from defects that would expose that substrate. To ensure this the coating thickness must be sufficient to prevent corrosion penetration within the required lifetime of the component. Alternatively, cathodic coatings may be used with exposure of the substrate provided that the substrate corrosion sites will rapidly passivate by the formation of insoluble corrosion products or that the rate of attack on the substrate is insufficient to affect adversely the service life of the article. Control can also be exercised by the use of multi-layer coating systems (for example, copper or nickel undercoats with gold coatings or nickel undercoats with chromium coatings), in which case the anodic/cathodic relationship of immediate importance is that between adjacent coating layers. However, as the period

Table 5.2 THICKNESS REQUIREMENTS FOR ZINC OR ALUMINIUM COATINGS ON STEEL

Environment	*Coating thickness* (μm) *for various service lives*		
	5 years	15 years	50 years
Outdoor industrial	30 Zn 50 Al	125 Zn 125 Al	— —
Outdoor rural	7 Zn 25 Al	30 Zn 37 Al	— —
Outdoor marine	15 Zn 50 Al	30 Zn 75 Al	— —
Indoor wet	15 Zn 50 Al	30 Zn 75 Al	— —
Indoor dry	7 Zn 37 Al	10 Zn 50 Al	15 Zn 75 Al

Abstracted from DD 24:1973 *Recommendations for methods of protection against corrosion on light section steel used in building*

of exposure of a composite system increases and corrosion penetrates through successive coating layers to the substrate, complex electrochemical relationships may be set up and one or more component of the system may be attacked at an enhanced rate.

Aluminium may be applied to steel by hot-dipping since the melting point of steel is sufficiently greater than that of aluminium, but if aluminium alloys have to be protected by pure aluminium coatings they must be applied by metal spraying or cladding. When chromium is to be applied as a coating metal, electrodeposition directly on to the substrate generally produces a coating with inadequate adhesion and/or inadequate protection of the substrate. With steel substrates nickel may be applied directly as an undercoat for chromium, but with zinc substrates an undercoat of copper must be applied beneath the nickel and with aluminium substrates it is necessary first to apply a chemical zincate treatment followed by copper and nickel undercoats. With plastics substrates it is first necessary to apply an electroless copper or nickel deposit in order to make the substrate conducting for electroplating; thick ductile undercoats are frequently necessary to ensure the retention of adhesion between the plastics substrate and the nickel and chromium layers when the effects of differential thermal expansion cause stress in the plated composites.

Effect of shape and size of articles

The shape and size of the article to be coated has little, if any, effect on the choice of coating metal, except in so far as economic considerations may limit the size of article which can be coated with a given, costly, material. However, shape and size influence considerably the choice of method by which the coating may be applied. Very small articles may be difficult or impossible to jig for normal electroplating; they may be coated by barrel plating techniques, by hot-dipping or perhaps by vacuum metallising. Similarly, excessively large articles may exceed the capacity of both electroplating or hot-dipping tanks — though some latitude is possible with the latter process by using double-end dipping techniques. In these cases the only practical solution is to apply the coatings by metal spraying techniques or to redesign the article in several smaller component parts that can be coated before assembly.

Intricately shaped components (particularly those with deeply recessed regions) are difficult to electroplate with coatings of even thickness because of the limited throwing power of plating solutions (see Chapter 3) — although some amelioration of this problem can be achieved, at additional cost, by using auxiliary and conforming-shape anodes to even out the current density distribution on the article being plated. Similarly, electrodeposits covering completely the inside of small-bore hollow sections can be obtained only by using internally placed anodes. Hot dipping may provide better coverage in these cases, although thickening of the coating in recessed areas may mar detailed shape and small-bore holes may become clogged with coating metal. Metal spraying techniques can cope well with irregularly shaped articles but coatings cannot be metal sprayed inside narrow bores. Chemical (electroless) plating, however, will coat the most complex shapes with even thickness both internally and externally. Probably the best way of handling both complex-shaped and excessively large components in order to achieve the best coatings is to redesign them so as to simplify the application of the chosen coating by the desired method. Indeed it is a fundamental principle in achieving both the best application and the best performance of coated metal components that the coating requirements be taken into account fully at the original design stage.

Effect of subsequent fabrication

Fabrication that must take place after application of any metal coating should always be considered when both the coating metal and its method of application are chosen. Obviously any cutting or trimming that has to be done after applying the coating will damage the coating and expose the substrate metal. Anodic coatings may well be able to cope with the exposed area of substrate by providing sacrificial protection, provided that the area in question is not too large, but the increased rate of consumption of the coating metal that results from the presence of exposed substrate may well markedly reduce the ultimate effective life of the coated article compared with that which would have been achieved if exposure had been avoided. In the case of cathodic coatings, however, any substrate metal exposed as a result of cutting after coating will itself be preferentially attacked; provision therefore has to be made to provide local protection in the exposed area or to repair the coating before the article is placed in service. The only coating process that can be readily applied *in situ* to a limited area of a large structure, thus repairing any damaged coatings, is metal spraying (although it may be possible with certain coating metals such as tin, lead and their alloys to effect localised repairs by soldering or brazing techniques).

Post-coating welding operations destroy the coating in the weld zone, and in part or perhaps the whole of the heat-affected zone, so local repair or protection is required similar to that necessary where cutting has taken place. In addition, the coating metal may well affect the welding process by alloying, causing unsound or embrittled weldments. A further hazard during welding can arise from the production of toxic vapours produced from the coating metals; for this reason cadmium should *never* be chosen as a coating metal for steel that must be subsequently welded.

Assembly of coated components may produce creviced regions, such as in bolted-up lap joints or beneath the heads of fasteners, and the susceptibility of the chosen coating metal to crevice corrosion must be borne in mind. Similarly, bi-metallic contact can occur on assembly; ideally, such contacts should be designed out of the structure, or assembly made with non-metallic (insulating) separators in the joint, but where these methods cannot be followed the coating metal chosen must be as compatible as possible with the dissimilar contacting metal. For example, where steel and aluminium must be in contact the steel should be coated with cadmium, since cadmium and aluminium when in contact do not lead to bi-metallic corrosion of the latter metal.

Mating and threaded components must be designed and produced so as to allow for dimensional changes occuring during coating, and the coating thickness and method of application must be chosen to achieve the best compromise between adequate fit and adequate protection with the minimum of post-coating machining.

The internal stress, ductility and brittleness of coating metals (and, where appropriate, of alloy layers) must be taken into account when choosing a coating metal and its method of application for a component that must be deformed during fabrication or in service. Electrodeposits such as chromium and some nickels can withstand only a small amount of deformation without cracking or spalling; the development of excessively thick alloy layers during hot-dipping also embrittles the coating and leads to failure on deformation. The hardness, ductility and frictional properties of a coating metal may be of considerable consequence in post-fabrication. A very soft coating such as lead, or to lesser extent aluminium, can deform readily under load; this may lead to more efficient elimination of some crevices but may also cause localised thinning of the coating or even exposure of the substrate. Sprayed zinc or aluminium coatings on steel are of especial value in applications where friction grip bolting is involved. Slip factors of the order of 0.45–0.55 are readily obtained with sprayed zinc coatings, and in the case of sprayed aluminium coatings the slip factor can rise as high as 0.7. Galvanised steel in the 'as galvanised' condition has a somewhat lower slip factor than sprayed zinc, owing to the smoothness of the deposit, but in service under dynamic loading a hysteresis loading cycle occurs which produces self-roughening of the faying surfaces with a consequent locking action which prevents any slip taking place. Conversely, the low torque resistance of cadmium makes it the best choice of coating metal for steel bolts for structures that must be assembled and dismantled frequently.

One point not directly concerned with fabrication but allied to it — and a matter that is frequently overlooked — is to ensure that all components in a composite structure have comparable effective service lives. Thus hot-dipped galvanised components having a coating thickness of some 50 μm or more may be assembled using fasteners that have been electroplated with zinc to a thickness of perhaps only 10–20 μm. In cases such as this the life of the fasteners will be only 20–40 per cent that of the rest of the structure (since the life of a zinc coating is proportional to its thickness) and unsightly rusting, or perhaps even collapse, will occur prematurely.

Mechanical factors

Mechanical factors that must be considered when choosing a coating are mostly those of stress during service — either dynamic or static. The application of heat during hot-dipping processes, and to a lesser extent during metal spraying, can adversely affect the mechanical properties of the substrate metal by partial or complete annealing during coating. If this occurs the strength of the completed component may be inadequate for its application, or the component may be distorted during coating so that subsequent assembly is difficult or even impossible.

When coated components are stressed during assembly or in service, failure can occur if the coating metal is susceptible to stress corrosion — for example, stressed copper or copper alloys exposed to ammoniacal environments. Alternatively, a substrate metal susceptible to stress corrosion may be completely protected by means of a suitable metal coating — for example, high-strength aluminium alloys coated with sprayed pure aluminium or with zinc. Dynamic stressing during service may produce flexing of a component, and in these cases if the coating is brittle it may crack and expose the substrate with consequent loss of protection; an example of this may be seen in the case of thick 'crack-free' chromium deposits, which fracture through brittleness when flexed (as in motor car bumper-bars or hub-discs), the cracks then propagating through the nickel undercoat to expose the steel substrate.

Coating by electrodeposition, by the nature of the process, frequently produces cathodic hydrogen at the metal surface, and this hydrogen may be absorbed in the coating and/or in the substrate. The presence of this hydrogen in certain metals can result in embrittlement, such as is the case with high-tensile steels, leading to brittle fracture when stressing occurs in service. Provision to made in the appropriate standard specifications to carry out stress-relief annealing treatments to remove or minimise these effects. Thus BS 1224:1970 (*Electroplated coatings of nickel and chromium*) specifies stress-relief annealing before plating for 1 hour at 130–210°C according to the type of steel, or after plating for 5 hours at 190–210°C or for 15 hours at 170°C if the higher treatment temperature would be harmful to the mechanical properties of the steel. Similar provisions are made for zinc or cadmium plated high-strength steel components, but in special cases it might be desirable to avoid coating by electrodeposition and use either metal sprayed zinc or vacuum metallised zinc or cadmium coating processes, thus avoiding exposure to hydrogen.

Sprayed metal coatings may also be preferable in applications where fatigue loading is involved, since the compressive stressing of the surface layers of the substrate by the grit-blasting pretreatment may improve fatigue properties. In applications where fretting of bolted joints can occur, the rough irregularities present on the surface of a sprayed metal coating can increase friction in the joint and so reduce fretting corrosion.

Metal-coated plastics materials are subjected to particular hazards through mechanical forces during service. The principal of these is the risk of rupture between the coating and the substrate through stresses arising during temperature changes because of the great disparity between the coefficients of thermal expansion of metals and plastics. In practice it may be necessary to incorporate a sufficient thickness of a ductile undercoat such as copper, which will prevent rupture when differential expansion and contraction occurs. Cases have also occurred in practice where plastics components plated with nickel and chromium have failed under stress in service by fracture in a sharply angled recess, although similar plastics components used unplated have performed satisfactorily. This mode of failure is caused by stress concentration in the notch of the recess cracking the chromium coating, the crack then propagating through the metal undercoats and through the plastics substrate. In such cases the only remedy lies in the redesign of the component to eliminate the notch effect.

Mechanical factors other than stress that must be considered are those involving movement. Relative movement between coated components and other components in an assembly occasions wear, and the wear resistance of the coating metal is then the primary consideration. In general the hardness of the metal coating is the principal factor of concern here, since normally a softer material wears when in moving contact with a harder material (although there are exceptions to this). Diffusion coatings and electrodeposited coatings of hard metals such as chromium and nickel are principally used for wear-resistant applications, but metal sprayed coatings subsequently machined and perhaps also heat-treated are often employed. A secondary property of metal coatings of importance for wear resistance is the adequacy of its adhesion to the substrate. Where adhesion is poor, rubbing action can cause localised rupture at the coating/substrate metal interface, leading to blistering or even complete spalling off of the coating metal.

Effect of environmental movement

Movement of the environment may also need to be considered. In applications involving exposure to moving liquids or gases erosion can occur. This may be insufficient to erode the metal itself but nevertheless sufficient to remove the protective films locally and thus set up anodic areas where enhanced corrosion will occur (for example, impingement corrosion of copper and its alloys immersed in moving water), or it may be of sufficent magnitude to damage mechanically the metal itself (as in cavitation corrosion). In either case premature localised penetration through a coating may occur, causing exposure of the substrate with consequent loss of protection, or even stripping the coating completely from considerable areas of the component as corrosion undercuts the coating and turbulence is increased in the moving liquid. The remedy in these cases lies in the selection of a coating material (e.g. nickel or nickel alloy) that offers improved resistance to erosion, or in the redesign of the component so that the erosive effects are reduced.

6

Testing coatings

In general terms the corrosion-control characteristics of a metal coating may be inferred and predicted from a knowledge of the performance of the metals concerned in the type of environment to which they are to be exposed. However, in practice, the full potential of a coating system can be realised only if the quality of the materials, their method of application and the necessary pretreatments of the substrate materials are properly controlled. Similarly, the performance of a given coating system of proven quality varies with minor changes in the local service environment, which may be very complex.

In consequence, one of the most important aspects of coating technology involves thorough and adequate testing of the coating quality to ensure compliance with specification requirements, and also corrosion testing in both naturally occurring and specially controlled corrosive environments so that a true optimum service performance can be accurately predicted and achieved in practice. Thus the testing of coatings may be seen to fall within two broad categories:

(a) quality control testing
(b) corrosion resistance testing

There is an interrelationship between these divisions, since the results of some quality control tests may indicate changes in expected corrosion resistance performance and the results of some corrosion tests may reveal variations in the quality of the tested coatings.

Quality-control testing

Tests for quality control may be divided into the following groups:

(a) visual inspection
(b) chemical composition

(c) thickness
(d) porosity
(e) adhesion
(f) stress
(g) ductility
(h) strength
(i) hardness
(j) wear resistance

In any given coating system or application the number of these tests that must be made varies; the method of test to be adopted also depends upon the materials concerned and the methods of coating application used.

Visual inspection

Testing by visual inspection might be thought to be a poor tool of very limited sensitivity, apart from meeting the requirement of aesthetic acceptability of a coated article. However, this method should not be despised or ignored since it can be a rapid and comparatively cheap means of detecting faults, and can provide a lot of useful information in the hands of experienced inspectors.

Gross defects of completely uncoated areas or mechanically damaged coatings are readily detected and rejected, and the reasons for the presence of these defects can often be seen. Thus an uncoated area may be seen by the shape or nature of the bare surface to be due to (say) physical shielding of the substrate from the coating metal during application, or the presence of surface contamination of the substrate. Damage to coatings may often be traced back to specific defects in handling procedures either during or after coating.

Apart from completely uncoated areas, cases where coating thickness varies with the geometry of the component can be detected by observation either of changes of coating-surface contours relative to the shape of the article or of the colour or reflectivity of the surface. For example, in the case of nickel + chromium electrodeposited coatings, the poor throwing power of the chromium plating bath can lead to thin, porous deposits in recesses and thick, dull deposits on asperities of the coated article, both of which are detectable by simple visual inspection. In extreme cases of lack of throw into recesses, nickel may remain exposed in these regions and can be detected by its hue, yellower than the bluish-white of a bright chromium deposit. Similarly, small skipped areas of a nickel coating can be detected by the pinkish colour of an exposed copper undercoat or the darker hue of the substrate metal itself.

Irregularities in the surface contour of a coated article may reveal defects in the substrate material or inadequate smoothing or polishing prior to coating. These substrate defects may also be revealed by blistering or gross lack of coating adhesion, though these latter defects may also be the result of inadequate pre-cleaning.

With hot-dipped coatings particles of dross may be entrapped in the solidifying coating metal as the article is removed from the molten bath. When this occurs the surface of the coated article is rough and nodular. The nodules may be bright with the colour of the zinc or aluminium coating metal or, if the dross itself protrudes through the surface, the nodules have a grey or blackened appearance. The presence of entrapped dross in hot-dipped coatings leads to inferior corrosion resistance in the affected areas.

Excessively rough or nodular deposits on metal sprayed coatings are generally caused by irregularity of atomisation of the coating metal in the spray gun, which results in the ejection of occasional molten globules larger than those normally produced or the production of particles that have not been fully atomised and melted. Although these rougher deposits have a less satisfactory appearance than the normal, smoother, coatings the nodules do not usually impair the corrosion resistance of the coating to any great extent.

Chemically deposited coatings may reveal powdery or discoloured areas, indicative of poor nucleation of the coating on the substrate material or of imbalance in the chemical composition or operating parameters of the electroless plating bath. In view of the fact that electroless deposits are usually thinner than many other types of coating, the presence of such defects should always be a cause of rejection. Since electroless coatings are often used as primary undercoats for other coating processes, careful inspection at the electroless deposition stage is of great importance.

With electrodeposited metal coatings the presence of visual defects, and their nature, can indicate the likely causes of their occurrence. Some defects will adversely affect corrosion resistance, while others may only affect the aesthetic appearance of the coated article. The exact causes of the occurrence of specific defects may be many and varied, and dependent upon the particular electrodeposition process employed. Detailed fault lists and methods of correction are published in the specialised plating handbooks and in suppliers' trade literature; these should be consulted in specific cases, but brief general guidelines may be summarised as follows:

(a) *Dullness or discoloration.* Indicative of lack of chemical balance of the plating bath or its contamination by foreign metals in solution. Dullness and discoloration do not seriously affect corrosion resistance, only aesthetic considerations.

(b) *Staining.* As distinct from discoloration of the plated deposits, staining is a purely superficial defect that is amost invariably due to inadequate rinsing and drying of the article after plating. Its occurrence may also be indicative of cracked or porous coatings and/or substrate metal, leading to entrapment of plating solutions in the voids and their subsequent oozing out on to the plated surface. Unlike discoloration, staining of a plated surface probably indicates reduced corrosion resistance since it is caused by the presence of highly corrosive plating salts, which are likely themselves to attack the metal surface itself.

(c) *Bare areas, pitted deposits or lack of adhesion.* These types of defect are clearly the most serious in their adverse effect on corrosion resistance. They may arise from a variety of causes, all of which are common to each type of defect, such as lack of chemical balance or incorrect operating parameters of the plating bath, contamination of the bath with dissolved foreign metals or insoluble foreign bodies suspended in the solution, and (most commonly) inadequate cleaning of the substrate metal prior to plating.

(d) *Nodular deposits.* The two most likely causes of nodular deposits are plating at excessive current densities and the presence of solid contaminants suspended in the plating solution. In the case of the former, the presence of nodules may not adversely affect corrosion resistance, except in so far as their presence can interfere with the continuity of thin coatings of other metals that may have to be deposited over them. If arising from the latter cause, however, the nodules themselves may well act as local anodic corrosion sites owing to their compositional difference from the bulk of the coating metal.

It will readily be seen from all the above that careful and informed visual inspection of metal-coated articles is a very worth-while operation that can provide considerable information of value in assessing the corrosion-resistant properties of the system as well as its aesthetic suitability.

Chemical composition

It is perhaps axiomatic that in order to achieve the correct performance of a metal coating it must conform to the correct chemical composition. However, in the case of most practical coating applications the testing of chemical composition is difficult or even impossible by direct means owing to the thinness of the coatings themselves and their intimate contact with other coating metals and metallic substrates.

For these reasons it is usually necessary to rely on analyses of specially prepared samples, which may be ladled from a hot-dipping vat, collected by metal spraying on to an inert substrate such as glass, or electrodeposited or chemically deposited on to a substrate under conditions that allow them to be freely stripped away after deposition.

Information on chemical composition can sometimes be obtained directly from a metal coating itself, either by the use of X-ray fluorescence or electron beam analysis of the metal surface, or by the examination of prepared metallographic cross-sections cut from the coated article (either by visual examination under the microscope or by using an electron-probe micro-analyser).

If coatings are thick enough it may be possible to machine or scrape away sufficient metal from their surfaces without contaminating the sample with either substrate metal or any substrate/coating alloy layers, to enable direct chemical analyses to be made.

Apart from the obvious necessity of ensuring the correct balance of elements in alloy deposits, examples of minor compositional variations on corrosion resistance that may be quoted are the sulphur content of electrodeposited dull, semi-bright and bright nickels and the presence of copper impurity in sprayed aluminium coatings (see Chapter 3).

Chemical spot tests may be carried out directly on the surface of a coated article for purposes of identification; three tests that are of particular value are as follows:

(a) *Spot test for identifying types of nickel.* It is possible to differentiate between dull, semi-bright and bright nickel electrodeposits by means of a chemical spot test using a solution made by dissolving 20 g chromic acid in 10 ml concentrated sulphuric acid and adding sufficient distilled water to make a total of 100 ml of solution. If a drop of this solution is placed on a freshly cleaned bright nickel surface it rapidly develops a dark brown coloration and the metal surface is attacked and darkened. On semi-bright nickel surfaces the reaction is slower and the colouring less definite, but with dull nickel the reaction is suppressed. In general, the darkening occurs within about five minutes, but in the case of bright nickels containing excessive amounts of brighteners (which often corrode more rapidly than usual in service) the reaction may take place much more quickly.

(b) *Spot test for the presence of a colourless chromate passivation coating on zinc or cadmium.* If a drop of 50 g/l lead acetate solution at pH 7.5 is placed on the surface, blackening occurs immediately if no passivation coating is present. Darkening will be delayed for five seconds for passivated coatings on cadmium and for 60 seconds on zinc.

(c) *Spot test for the presence of hexavalent chromium in a passivated coating.* A solution is prepared containing 0.4 g diphenylcarbazide dissolved in 20 ml acetone and 20 ml ethanol; 20 ml of 75 per cent phosphoric acid and 20 ml of water are then added. If a drop of this solution is applied to a passivated coating, the presence of hexavalent chromium is revealed by the development of a red coloration after a few minutes.

Thickness testing

Since the performance of all metal coatings depends upon coverage of the substrate with a complete coating of adequate thickness, and furthermore the life of the coating is dependent upon its thickness unless it is completely inert to the corrosive environment, it follows that one of the most important tests is the determination of coating thickness.

In cases where the coating is of sufficient thickness and the dimensions of the uncoated component are accurately known, it is obviously possible to determine the coating thickness directly by use of conventional measuring instruments. However, this method can generally be used only in the case of heavy-duty engineering coatings, since with the majority of coatings applied for normal corrosion-control purposes their thickness is small in relation to the dimensions of the uncoated article and so insufficient accuracy can be achieved even if direct measurement is feasible.

Ideally, coating thickness measurements should be by non-destructive means so that the coated article remains undamaged and can be put into service after testing; again this is not always possible, however, and where destructive testing methods must be employed it is necessary either to test samples from a batch of components or to effect repair of local damage produced by thickness testing. In either case, sufficient determinations must be made in appropriate areas to ensure that the figures obtained allow for inevitable thickness variations over different regions of the surface of a coated article and between nominally replicate articles.

Methods of determining coating thickness and their general applicability are summarised in *Table 6.1.* More detailed consideration of these methods follows.

Magnetic methods

The thickness of non-magnetic coatings on magnetic substrates or of

Table 6.1 METHODS OF THICKNESS TESTING

Type of test	*Test method*	*General applicability*
Non-destructive	Magnetic	All coatings on ferrous substrates and some nickel coatings
	Eddy current	Metal coatings on non-metallic substrates
	X-ray spectrometry	All systems
	β back-scatter	Almost all systems provided that the difference in atomic number between coating and substrate exceeds 5
May be either destructive or non-destructive	Light section microscope	All systems
	Interferometry	All systems
	Profilometry	All systems
Destructive	Chemical dissolution.	All systems
	Coulometric	Almost all systems except coatings of precious metals
	Microscope cross-section	All systems

electrodeposited nickel coatings on either magnetic or non-magnetic substrates can be determined by using an instrument, containing a permanent magnet or an electro-magnet, that measures either the force required to overcome the magnetic attraction between the magnet and the metal as influenced by the coating, or the reluctance of the magnetic flux path passing through the metal composite. In use, it is necessary to calibrate the instrument using standards of known thickness, which should be of the same coating/substrate combination as the materials under test.

Accuracy of determination is affected by the geometry of the coated surface, by the design and method of application of the magnetic probe, and by the thickness and magnetic properties of the substrate metal. Except on thin coatings (generally less than 5 μm) the normal accuracy achieved is ± 10 per cent, the maximum measurable coating thickness being governed by the available strength of the magnetic field. These methods of thickness testing are covered by two International Standards, ISO 2178 and ISO 2361.

Eddy current method

This method utilises an instrument in which eddy currents caused by the differences in electrical conductivity between the coating metal and the substrate material are measured. The method is particularly suited to the measurement of the thickness of metal coatings on non-metallic substrates or of non-metallic coatings on metallic substrates (for example, anodic oxide coatings on aluminium, or paint or lacquer films on metals generally) and gives an accuracy of better than ± 10 per cent. It may be used, with caution, for particular all-metal composites where the electrical conductivities of the coating and substrate metals are sufficiently different, but great care is required in these applications. Calibration, in all cases, is by means of standards of known thickness.

Somewhat akin to the eddy current method is the thermo-electric method. A heated probe applied to the coating surface generates thermo-electric currents at the differential metal interface, and these currents may be measured by suitable instrumentation calibrated against standards of known thickness. Although attempts have been made to produce practical instruments employing this method of determination the results achieved have been found to be highly sensitive to the constructional details of the probe assembly, to temperature variations in the test pieces, and to minor compositional differences in the metals in the composite. For these reasons, accuracy and reproducibility of results are not good and the practical use of this type of instrument is very limited.

X-ray spectrometry

If a metallic composite is irradiated by X-rays, secondary radiation is produced of wavelengths characteristic of the elements present in both the coating and substrate metals. By use of a single-crystal spectrometer, a particular radiation wavelength characteristic of either metal can be selected for individual intensity measurements by electronic pulse-counting techniques. The measured intensity is related to the coating thickness, up to a limiting value of thickness.

Two methods of measurement may be employed. In the first the intensity of emission of secondary radiation characteristic of the coating metal is measured; this measurement increases with the coating thickness up to a limiting thickness, although a small amount of radiation will be detected caused by scattered background radiation from the bare substrate metal. In the second method the intensity of emission of secondary radiation characteristic of the substrate metal is measured; this decreases with increasing coating thickness (owing to absorption of the radiation by the coating metal) until the limiting thickness is reached, after which only a constant, minimum, scattered background radiation from the coating metal will be detected. In both methods the limiting thickness that may be measured is dependent on the atomic number of the coating metal, and the intensity/thickness curves are asymptotic to the thickness axis. Calibration must be carried out by the use of standards of known thickness with the same coating/substrate combination as the material under test.

There are many factors affecting the accuracy of the determinations, but if these are carefully controlled an accuracy of ± 10 per cent can be achieved over the major portion of the limiting thickness range. As already mentioned, this limiting thickness varies with the coating metal; to quote an example the limiting thickness for nickel is about 10 μm and the range of maximum accuracy is 0.25–7.5 μm. All coating/substrate combinations can be tested, but not multi-coat systems. The standard method of thickness testing by X-ray spectrometry is described in ASTM Test Method B568-72.

Beta back-scatter method

This is again a radiation method of measuring coating thickness using an instrument in which a radio-isotope emits beta-rays, which are reflected back by the atoms of the coating metal. The intensity of the back-scattered beta-rays varies with the coating thickness and with the atomic number of the coating metal, which also governs the maximum thickness that can be measured. The intensity of the back-scattered beta-rays is measured with a pulse counter and the thickness is then obtained from a graph of intensity versus thickness. The curve on the graph is linear up to a certain thickness of coating, is logarithmic over the major thickness range, and becomes hyperbolic as the saturation thickness is reached. The saturation thickness increases with reduction in the atomic number of the coating metal, ranging from 50 μm for a metal of high atomic number (such as gold) to 300 μm for metals of low atomic number (such as copper or nickel).

As with other radiation methods the accuracy of measurement can be affected by the atomic relationship between coating and substrate (an atomic number difference of at least 5 is necessary for the successful operation of the method) and by the substrate thickness or the presence of thin intermediate coatings of different composition. In the case of multi-coat systems where the atomic numbers of the different coating layers are similar (e.g. copper + nickel + chromium), the beta back-scatter method measures only the total thickness of the composite without differentiating between its component parts. Instrumental and geometric variables, which can be controlled, also affect accuracy, and calibration of the instrument must be carried out on standards of similar composition to the work being tested. The accuracy that can be achieved with this method is ± 10 per cent.

Light section microscope method

By directing a beam of monochromatic light through a microscope objective on to a reflecting plane metal surface at an angle of 45°, a reflected line image may be viewed with another objective lens. If the surface is not plane the light rays are deflected by an amount proportional to the surface irregularity so that, if a small area of a metal coating within the illuminated beam is removed so as to expose the substrate surface, the deflection of the beam provides an absolute measure of the coating thickness. In the case of transparent coatings (i.e. non-metallic coatings such as clear anodic oxide coatings on aluminium) reflections are obtained from both the coating and substrate surfaces without removal of the coating, so the method is non-destructive. With opaque coatings (i.e. metallic coatings or coloured non-metallic coatings) removal of the coating in the illuminated area must be effected; thus the method is destructive to at least some degree. The method is applicable to any metallic coating, provided always that it can be stripped from the substrate without damage to the latter, and is applicable to coating thicknesses up to about 5 μm. The accuracy of measurement is governed by the sensitivity of adjustment of the optical system for reading the deflection of the light beam, but ± 10 per cent should be readily obtainable.

Interferometry

This is an optical method employing a beam of monochromatic light directed on to a step between coating and substrate, in a similar fashion to that of the light section microscope. However, instead of measuring the deflection of a reflected beam the microscope technique is used to count the number of interference fringes produced by light scatter resulting from the step height of the coating. The number of fringes multiplied by half the wavelength of the light employed equals the coating thickness.

Two variations of the basic method may be employed:

(a) The double-beam technique, which is suitable generally for coatings in the thickness range 0.3–10.0 μm.

(b) The Fizeau multiple-beam technique, for coatings thinner than those that can be measured by the double beam technique; this method is suitable for coatings in the thickness range 0.02–2.0 μm.

During the production of interference fringes the reflected light undergoes a phase change, the magnitude of which varies with the substance producing the reflection; inaccuracies may therefore occur owing to the differences in phase shift produced by coating and substrate respectively. Generally these inaccuracies are small in relation to the coating thickness, particularly for the thicker types of coating, but they may be eliminated by applying a very thin reflective coating of a single vacuum-deposited metal on both coating and substrate prior to measurement. By this means, and the use of the multiple-beam technique, it is possible to achieve an accuracy of better than ± 0.01 μm. The applicability of the method and its non-destructive or destructive character are the same as for the light section microscope method.

Profilometry

As with the two optical methods just described, it is necessary to obtain a step between coating and substrate by local removal of the coating when using profilometric methods of thickness determination. In these cases, however, the thickness is measured by recording the extent of movement of a stylus that follows the profile of the step as it is drawn across the test surface.

Electronic instrumentation is employed to amplify the movement of the stylus and produce a graphical trace of the amplified profile from which direct measurement can be made. The method is applicable to all coating systems provided that a step can be produced without damage to the substrate. Thicknesses in the range 0.005 μm to 1 mm can be measured with an accuracy of better than ± 10 per cent.

Chemical dissolution methods

Chemical dissolution methods of determining thickness fall into four categories:

(a) Dissolving the coating from a small area by applying the solution at a controlled rate and measuring the time to penetrate to the substrate (BNF jet test).

(b) Dissolving the coating completely from a weighed and measured sample and noting the weight loss (gravimetric method).

(c) Dissolving the coating completely from a known area of sample and determining the amount of coating metal in solution in the reagent used for dissolution (analytical method).

(d) Treating a sample of the coated article with a reagent that evolves a gas during the reaction, and noting the time of gassing (gassing tests).

BNF jet test. The apparatus for carrying out this test (*Figure 6.1*) consists of a glass capillary jet of controlled dimensions, attached through a stopcock to a glass container for the test solution. The container, fitted with a constant-solution head device, is secured so that the jet is suspended vertically above the surface of the test specimen, which is inclined at an angle of 45°. The solution temperature must be

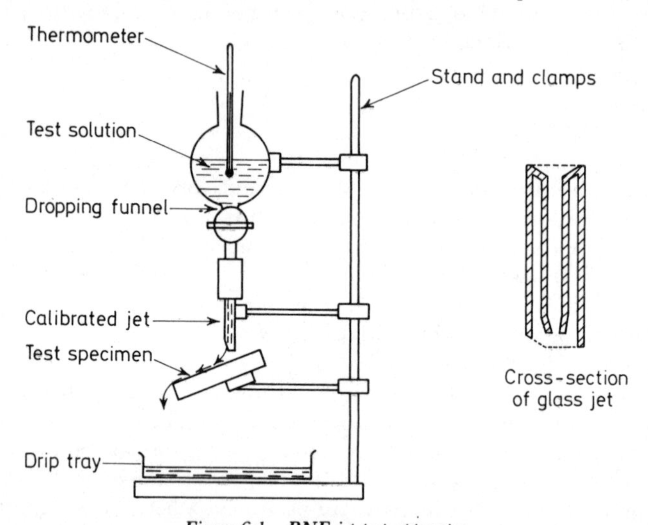

Figure 6.1 BNF jet test apparatus

recorded as it flows freely from the jet under the pressure of the constant head, and the surface of the test specimen is observed for penetration of the coating in the area where the jet impinges. Penetration is detected by the appearance of a colour change either due to the actual colour of the underlying metal (e.g. copper) or induced by the reaction of the test solution with the underlying metal (e.g. blackening of aluminium or zinc alloy substrates). The time of jet flow to penetration is noted and the coating thickness calculated by reference to graphs that show the time to penetrate 2.5 μm of coating over a range of solution temperatures.

The test is applicable to cadmium, cobalt, copper or bronze, lead, nickel, silver, tin or tin–zinc, and zinc electrodeposits on aluminium, copper or brass, steel and zinc substrates. Where multi-coat systems are involved it is possible to determine the thickness of the individual coating layers by successively operating the jet with the appropriate solution on the same area of the specimen surface. The time required to determine the thickness of an individual coating layer is one or two minutes, and the general accuracy is ± 15 per cent. However, the dissolution reaction is sensitive to the purity of the coating metal, so it is not applicable to coatings that progressively alloy with the substrate; special difficulties occur in obtaining accurate results with coatings such as organic bright nickels, owing to the effect of varied levels of brightener concentration on the rate of dissolution.

Test solutions used consist of ferric chloride/copper sulphate for cobalt, copper or nickel coatings; acidified ammonium nitrate for cadmium or zinc coatings; potassium iodide/iodine solution for silver coatings; trichloro-acetic acid for tin coatings and acetic acid/hydrogen peroxide solution for lead coatings.

An advantage of this method is the ability to determine the actual coating thickness in any selected position on a plated surface where the jet can conveniently be directed, as opposed to other chemical dissolution methods which determine only the average coating thickness over the whole area from which the coating has been dissolved (see the BNF jet test booklet[3]).

Gravimetric method. Gravimetric (or 'strip and weigh') tests are used for a wide variety of metal coatings and can give an accuracy of ± 5 per cent. However, they have the disadvantage of being completely destructive and of providing only the average coating thickness over the whole area tested, so they give no indication of local variations in coating thickness. The principle is the simple one of weighing a sample of known surface area both before and after the coating is removed by immersion in a suitable chemical solution that will attack the coating metal. The solution may be one that does not itself attack the substrate metal, or may contain a suitable inhibitor that either prevents attack on the substrate metal or reduces it to a very small amount, which can be calculated and deducted from the weight loss as a blank. The loss in weight due to the removal of the coating is converted to thickness by dividing by the product of the surface area tested and the density of the coating metal.

Cadmium, tin or zinc coatings may be stripped from steel substrates by use of a hydrochloric acid solution containing antimony trioxide or trichloride, which act as inhibitors to prevent the acid attacking the steel (see British Standards 1706 and 1872). Alternatively, cadmium may be stripped in a 30 per cent solution of ammonium nitrate and zinc in a solution of 5 g ammonium persulphate and 10 ml ammonium hydroxide in 90 ml water (see British

Standard 3382). Tin–nickel alloy coatings are stripped electrolytically in a solution containing 20 g/l caustic soda and 30 g/l sodium cyanide or, when coated on copper, by immersion in concentrated phosphoric acid (see British Standard 3597). Silver coatings are stripped by first immersing in a 1/19 volume mixture of concentrated nitric and sulphuric acids until blackening occurs, after which they are transferred to a 250 g/l solution of chromium trioxide in concentrated sulphuric acid (see British Standard 2816). In the case of gold coatings the procedure is to strip the substrate from the coating by dissolution in concentrated nitric acid, subsequently filtering off the unattacked gold which can then be washed, dried and weighed (see British Standard 4292).

Analytical method. In analytical methods the coating is chemically stripped from a measured area of substrate and the amount of metal in an aliquot of the solution is determined by a suitable analytical technique. Copper coatings may be stripped in a solution containing 10 g ammonium persulphate, 100 ml ammonium hydroxide and 100 ml water and the copper content of the solution determined colorimetrically (see British Standard 3597). Silver coatings may be stripped in a concentrated sulphuric/nitric acid mixture and the silver content of the solution determined by titration against ammonium thiocyanate solution (see British Standard 4290).

Gassing tests. It is possible to utilise the generation of hydrogen gas by chemical reaction of a coating metal with an acid to determine the coating thickness. Thus chromium evolves hydrogen when attacked by hydrochloric acid, and the thickness of a chromium coating can be calculated by measuring the time of gassing in hydrochloric acid; a 0.25 μm thick coating evolves gas for approximately 10 seconds at 20°C (see the BNF jet test booklet[3]). Similarly, a cadmium coating 2.5 μm thick gasses for 5 seconds when immersed at 20°C in a solution of 10 g nickel sulphate in 100 ml hydrochloric acid (see British Standard 1706). With both these reactions the gassing time must be corrected for temperature variations above or below 20°C.

In the case of zinc coated steel wire the average coating thickness may be determined by stripping the zinc from a known length of wire of known diameter by immersion in standard hydrochloric acid solution inhibited with antimony trioxide or trichloride and collecting the evolved hydrogen; if this is done the volume of gas (in ml corrected for temperature and pressure) divided by the product of the length and diameter of the wire sample and multiplied by a constant (872) gives the coating weight per unit area (g/m^2) (see British Standard 443).

Coulometric method

The principle of the coulometric method of thickness determination is the reverse of that of electrodeposition — namely, anodic dissolution of metal over a known area with a measurement of the quantity of electrical charge consumed in the process. From a knowledge of the area over which electrolysis has taken place and the Faradaic electrochemical equivalent of the metal concerned a simple calculation will convert the number of coulombs of electricity consumed in the process into the thickness of coating metal dissolved. In order to achieve accurate results with this calculation it is necessary to ensure that dissolution is achieved at a known, constant anodic efficiency (preferably 100 per cent); the electrolyte must be chosen to ensure that passivation or excessive polarisation effects do not occur, and furthermore that the chosen electrolyte does not chemically attack the coating metal in the absence of the electric current. It is also, of course, essential to define accurately the anodic area.

It is possible to utilise the coulometric method under conditions of electrolysis that do not provide 100 per cent anodic efficiency, but in this case it is necessary to know the anodic efficiency accurately and to ensure that it remains constant if the calibration is to be made using a simple calculation involving the electrochemical equivalent. If these conditions are not fulfilled it is necessary to calibrate the instrument by operating on standard specimens with known thicknesses of the relevant coating metals. The American producers of the Kocour Electronic Thickness Tester normally recommend the use of standard specimens for calibration rather than the use of theoretical calculations.

In a typical instrument employing 100 per cent anodic dissolution efficiency[4] a constant current of 80 mA is employed in the cell at a voltage in the range 1.5–3.5 chosen according to the metal being dissolved. The voltage is adjusted to be marginally greater than that at which dissolution may be maintained and remains constant until all the coating metal has been dissolved, at which time changes in the electrode process occurring as a result of exposing the (different) underlying material cause a variation in the applied cell voltage; this indicates the end point of the determination (by tripping a cut-out relay). An integrating coulometer in series with the cell records the number of coulombs consumed during the dissolution reaction; this figure multiplied by a constant for the coating metal concerned enables the coating thickness to be directly calculated. (Later models of the instrument replace the integrating meter with a direct readout display of thickness in arbitrary units based on the accurate measurement of the time that current is passed during the determination coupled with very accurately controlled constant-current conditions during testing.) The cell used consists of a stainless steel tube about 25 mm in diameter × 40 mm long with a flexible plastics base containing a central circular aperture 5 mm in diameter. The stainless steel cell wall forms the cathode, and the workpiece is connected electrically to the instrument to form the anode.

By retaining the cell in position and successively using different electrolyte solutions appropriate to the relevant coating metals, the thickness of each layer of a composite coating system can be determined.

The thickness of coatings of cadmium, chromium, copper, lead, nickel, silver, tin and zinc may be determined on a wide range of substrates, including plastics materials. The accuracy of the method is better than ± 10 per cent over a range of coating thicknesses from 0.2 μm to 50 μm; it is also possible to use the method to measure greater coating thicknesses such as those used for hard chromium coatings for engineering applications, but in these cases frequent replenishment of the electrolyte in the cell is necessary and some inaccuracies may be introduced owing to the effects of the walls of the coating metal surrounding the dissolved area on the electrolysis reaction.

The method is, of course, destructive over the small area covered by the cell aperture, but thicknesses may be determined at any chosen point on the surface of an article provided only that its geometric shape allows the cell to be applied in the desired position. Individual thickness determinations take only a few minutes' operation of the instrument. The coulometric method of thickness testing is described in International Standard ISO 2177.

Microscope cross-section method

Direct measurement of local coating thickness by means of examination under a microscope of a mounted and polished cross-section cut from the coated article is the method that is universally applicable, irrespective of the materials concerned and of the shape of the coated article. Using this method it is also possible to determine accurately the extent of any alloying between coating and substrate.

Since only direct observations and measurements are involved this method is most frequently specified as the referee method in cases of dispute, and is also frequently used to check the accuracy of non-destructive methods of determining coating thickness. Using normal mounting and polishing techniques with conventional optical microscopes thickness can be readily measured with an accuracy of ± 1μm, but by employing the taper-section method of mounting the specimens for examination accurate measurements can be made of coating thicknesses in the range 0.1–1.0 μm. Further extensions of the method can be made by using electron microscope techniques in order to measure even thinner deposits.

Care must always be taken when preparing the micro-sections to avoid breaking up the coating during cutting or deforming it during mounting, and to obtain the true edge profile during polishing. It is often advantageous to back-up the coating to be measured by protective over-coating with some different metal before preparation of the micro-section; this applies particularly to the case of very thin deposits that require taper sectioning techniques or with brittle coatings that may be chipped during preparation.

As well as determining thickness, the microscope cross-section method is of great use for obtaining many additional pieces of information about metal coatings. The extent of alloying with the substrate has already been mentioned, and the metallurgical structure of the coating metal, including the presence of porosity or inclusions, can be readily studied. Such metallurgical data may enable more exact identification of types of coatings to be made; for example the crack pattern of micro-cracked chromium and the minute pores in micro-porous chromium differentiate these deposits from conventional or 'crack-free' deposits, and the lamellar microstructure of organic bright nickels is in contrast with the columnar crystal structure of dull nickel deposits. Evidence may be found of stress in deposits and of areas of defective adhesion; coating hardness can be measured by micro-hardness techniques. The examination of coated articles after corrosion testing or service may enable performance to be more readily evaluated or the causes of failure to be established.

Porosity testing

The presence of pores in a metal coating can markedly influence its corrosion protective value; generally pores are deleterious to performance, but in certain applications (e.g. micro-cracked or micro-porous chromium coatings) a satisfactory pattern of porosity is essential to a correct functioning of the protective system. In either case a test to reveal the pattern of discontinuities in a coating is a useful tool in quality control. Several standard tests exist; most of them are basically forms of accelerated corrosion tests, which reveal pores by the production of coloured corrosion products of the underlying metal layers at the sites where these layers are exposed by the coating discontinuities. However, the use of porosity tests should always be approached with caution since the presence of substrate corrosion products after testing does not necessarily prove that an open pore existed prior to testing; locally thin coating regions that might be adequate to withstand the corrosive environment encountered in service may be penetrated by the action of the specific, aggressive corrodent used in the porosity test. The situation may be summed up by saying that it is often an open question whether the porosity test reveals pores or creates them.

The following porosity tests are in common use:

(a) *The Preece test.* Immersion of a coated steel component for one minute in a solution containing 360 g/l copper sulphate that has been neutralised with copper hydroxide and filtered will reveal discontinuities in the coating by depositing metallic copper on them but not on the remaining coated surface. The method is specified in British Standard 443 as an acceptance test for galvanised steel wire.

(b) *Ammonium persulphate test.* If tinned copper articles are immersed for 10 minutes in a 10 g/l solution of ammonium persulphate containing 20 ml/l ammonium hydroxide, discontinuities in the tin coating can be detected by the appearance of the dark blue cupro-ammonium colour complex in the solution in their vicinity. The test can be made quantitative by estimating the copper content of the solution colorimetrically after test.

(c) *Ferroxyl test* (see British Standard 4758). Porosity in nickel-coated steel can be detected by this test. Special test papers are prepared by treating filter paper by immersion in a solution containing 50 g/l sodium chloride and 50 g/l gelatine, followed by drying. Before use the papers are re-wetted in a 50 g/l sodium chloride solution containing a little wetting agent, and then squeegeed on to the nickel-coated article and left for 10 minutes. After removal the papers are immersed in a 10 g/l solution of potassium ferricyanide; blue marks develop on the paper in the regions where the steel was exposed through discontinuities in the nickel.

(d) *Electrographic tests.* As with the ferroxyl test, papers impregnated with solutions containing specific colour reagents are employed for electrographic tests. In these tests, however, the reagent solutions in the papers that are in contact with the coated test-piece are used as electrolytes. The test-pieces are made anodic and the wet papers are backed up by a metal cathode; basis-metal cations pass through the pores in the coating metal and react with the colour reagent in the paper to produce coloured spots, which provide a permanent map of the coating surface. An electrographic test described in British Standard 4025 uses an aqueous solution of cadmium sulphide for detecting porosity in coatings applied to copper substrates, brown copper sulphide spots being formed. Similarly, a solution of dimethylglyoxime in alcohol can be used to reveal discontinuities in coatings on nickel substrates, and a solution of rubeanic acid in alcohol to reveal porosity in gold deposits on copper or nickel substrates.

(e) *Dubpernell test* (see British Standard 1224). For revealing the pattern of micro-cracks or micro-pores in special micro-discontinuous chromium deposits, the chromium plated article is electroplated in an acid copper sulphate solution (200 g/l $CuSO_4$ + 20 g/l H_2SO_4) at room temperature for one minute at a current density of 30 A/m^2. Copper is deposited on the chromium surface only where cracks or pores are present, and the coppery pattern can then be examined under a microscope. The test should preferably be made immediately after plating since copper deposition may be inhibited if the plated article is stored for any appreciable length of time; in such cases the chromium surface should be reactivated prior to copper deposition by a four-minute immersion in a 10–20 g/l nitric acid solution at 95°C and thoroughly rinsed.

(f) *Sulphur dioxide tests.* Any of the sulphur-dioxide accelerated corrosion tests (see pages 167–168) will reveal coating discontinuities in gold or chromium deposits by corrosion of the underlying metal, but generally these tests are so searching that the discoloration due to porosity is masked by the much greater quantity of corrosion produced by heavy attack of the aggressive reagent on the underlying metal. For this reason specialised porosity tests utilise environments containing very much smaller quantities of sulphur dioxide so that pore sites are not seriously enlarged and the spread of staining by substrate corrosion products is restricted.

A special, rapid sulphur dioxide porosity test for gold coatings on copper, silver or nickel substrates is described by Clarke and Sansum[5]. Specimens are exposed for two hours in a closed glass vessel in which sulphur dioxide is generated from a solution made up of 4 vols 20 per cent w/v sodium thiosulphate solution + 1 vol of 50/50 sulphuric acid. The volume of solution used should be 1/40 the volume of the container and the test chamber maintained at 60°C. Both reagents and specimens should be preheated to the test temperature before insertion in the chamber so as to avoid surface condensation, which would obscure the pattern of pores revealed by the action of the sulphur dioxide.

Adhesion testing

With most properly-applied metal coatings the chemical or metallurgical nature of the bond between coating and substrate is of such high strength that lack of adhesion in service is unlikely to occur. Exceptions to this general condition arise in the case of sprayed metal coatings, where the bond is a purely physical one relying on the mechanical keying action between the roughened substrate and the sprayed metal; also with metal coatings on plastics substrates, where a weak physico/chemical bond between the plastics substrate and the deposited metal is involved; and also in some chemically deposited metal coatings and most chemical passivation coatings, where only a weak chemical bond is produced.

However, the adhesion of any metal coating to its substrate may be seriously impaired by incorrect operation of either pretreatment or coating processes, and adhesion tests are necessary to detect such processing irregularities or to measure the limited bond strength of the special cases quoted. Because of the practical difficulties of measuring adhesion most of the test methods are empirical and operate on the 'go, no go' principle; for this reason many of them rank as non-destructive tests provided that the adhesion of the coating is adequate to withstand the action of the test, and they only become destructive tests when performed on samples that have inadequate coating adhesion. The following tests may be used:

(a) *Burnishing tests.* If the surface of a smooth metal coating is burnished by rubbing with a smooth, rounded and polished steel, agate or bone tool, heat generated by friction can produce blisters in regions where adhesion between coating and substrate is inadequate. Tests of this type are called up in BS and ASTM specifications for coatings of cadmium, gold, silver, tin, tin–nickel and zinc. Somewhat akin to these burnishing tests for metal coatings, the adhesion test for chromate passivation coatings on zinc and cadmium coatings consists of rubbing the surface of the chromate coating with either a white india-rubber or tissue paper; inadequate adhesion is revealed by the presence of a yellow stain on the rubber or paper.

(b) *Bend tests.* Bend tests may be used both for adhesion testing and for testing the ductility of a coating. For either purpose the same testing procedure is used, namely deforming a test sample round a mandrel of specified curvature; the difference between the two forms of the test lies in the criterion adopted for assessing failure. When testing for ductility cracking in the cross-section of the coating is required, while for adhesion testing failure is indicated by lifting of the coating from the substrate. In British Standard 443 adhesion of galvanised coatings on steel wire must survive close coiling round a mandrel whose diameter is four or five times that of the test wire, and in British Standard 2816 plated silver coatings must survive 3 × 90° reverse bends round a 4 mm radius.

(c) *Quench tests.* Nickel, tin and tin–nickel coatings may be tested for adequate adhesion by heating to a temperature of 150–350°C (according to the nature of the substrate) and quenching in water without failure (see British Standards 1224, 1872 and 3597).

(d) *Scribe tests.* Satisfactory adhesion of a coating to its substrate may be judged by its ability to withstand the scribing of lines cutting through its thickness down to the substrate without the coating flaking away. This method of testing is specified in British Standard 2569 for sprayed zinc or aluminium coatings, and no break is allowed between two parallel lines scribed a distance of ten times the coating thickness apart. In British Standard 4292 a scribe pattern of 2 mm squares is specified, flaking within any of these squares being indicative of inadequate adhesion.

(e) *Pull-off tests.* The adhesion of sprayed metal coatings may also be tested by means of a tensile pull-off technique. A cylindrical dolly with a plane face is secured to the surface of the coating with a suitable adhesive so that the axis of the dolly is normal to the coating surface. The coating is then trepanned closely around the edge of the dolly and a tensile pull applied to produce failure (*Figure 6.2*). The test may be carried out in two

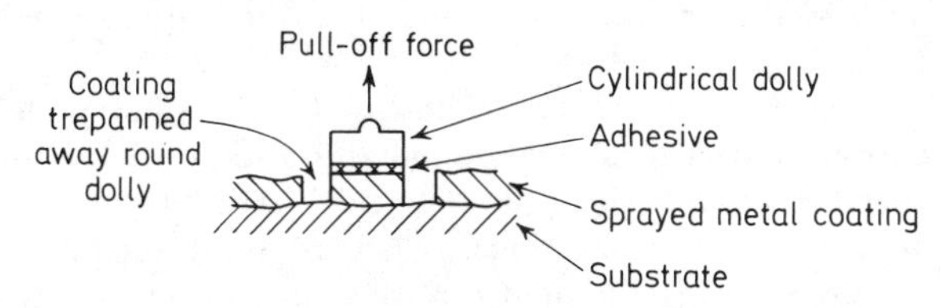

Figure 6.2 Method of testing adhesion of sprayed metal coatings

ways, either by using two dollies secured co-axially on opposing sides of the coated test piece and gripping them in the opposing jaws of a tensile machine, or by using only one dolly and securing the test piece rigidly while the dolly is pulled off. The method gives a quantitative figure for adhesion when failure occurs at the coating/substrate interface, or a minimum adhesion value if failure occurs within the coating or the substrate body. However, the accuracy of the method is poor because of the difficulty in securing the dolly so that the tensile pull is exactly normal to the coating surface, and furthermore because of the effect of the adhesive, which may penetrate through the porous sprayed metal coating and so increase the apparent bonding to the substrate.

(f) *Peel test.* The peel test is somewhat similar to the pull-off test in that a tensile pull normal to the coated surface is employed. The method is specific to plated metal coatings on plastics substrates, and must be performed on specially prepared test pieces with plane, flat surfaces to which a thick ductile copper coating is applied after the electroless metal deposit has been obtained on the plastics substrate. The purpose of the test is to measure the bond between the electroless metal deposit and the plastics substrate, because this bond is sensitive to the pretreatment processes applied to the plastics material and also to the physical condition of the plastics material. Two parallel lines are scribed 25 mm apart (or some other suitable distance apart) so that they cut through the electrodeposited copper layer (which has been applied 15 μm thick) and through the electroless metal layer, down into the underlying plastics material. A 'tongue' of metal between the scribed lines is lifted up by inserting a blade between the plating and the substrate at the edge of the specimen and this tongue is secured to the jaw of a tensile machine and the specimen secured rigidly. The load required to peel off the metal from the plastics substrate is recorded as the 'peel value', care being taken during pulling to maintain the tensile pull at 90° to the surface of the specimen by means of suitable linkages in the testing machine. The method is fully described in ASTM Method B533-70.

(g) *File test.* A rough method of testing for adhesion of nickel plus chromium coatings on both metallic and plastics substrates is specified in British Standards 1224 and 4601. It consists of applying a file at an angle of 45° to a cut edge on a plated article, drawing the file from the substrate across the coating and observing any signs of lifting of the coating from the substrate. Such lifting must not occur if the adhesion of the coating is to be considered satisfactory for passing the test.

(h) *Thermal cycling test.* This test is also specific to plated plastics articles, and is used to establish whether the bond between the coating and the substrate will withstand the stresses induced by differential thermal expansion and contraction between the metal and the plastics when temperature fluctuations occur in service. Plated articles, which should be selected entire from production, are first cooled to −40°C for one hour, then allowed to regain room temperature for one hour, after which they are heated to +80°C for a further one hour and cooled again to room temperature. Four complete testing cycles are used and the articles are examined for visible defects at each return to room temperature during each cycle. The acceptance requirement laid down in British Standard 4601 is a complete absence of visible defects such as cracking, blistering or peeling throughout the test; for less severe service conditions the cooling portions of the test cycles are omitted and only room temperature and +80°C test temperatures are used. As with bend tests, adhesion is not the only variable influencing the test results but rather, probably, a combination of adhesion, stress, ductility and strength. In many ways the thermal cycling test can be considered superior to the peel test, even though it provides only a qualitative measure of adhesion while the peel test puts a numerical value on the adhesion. This is because the peel test must be carried out on a specially prepared test piece that has been specially plated with a coating that bears little or no resemblance to the coatings applied to production items; furthermore there is no guarantee that the plating conditions remain unchanged as between peel test sample and actual production items. It has been found that the thermal cycling test is somewhat more searching than service, in that an article that fails the thermal cycling test may not suffer loss of adhesion in service temperature fluctuations but on the other hand success in passing the test appears to indicate with 100 per cent reliability that adhesion loss will not occur in service. The thermal cycling test therefore represents a very useful method of positive quality control. With peel testing, however, minimum peel values are set for acceptance, but many cases have occurred where service failure has occurred even though the minimum peel value has been well exceeded and similarly successful service has been obtained from articles with lower than minimal peel values.

Stress testing

Testing the internal stress of metal coatings can only be achieved in one of two ways. In the first of these it is necessary to strip the coating complete from its substrate by either chemical or mechanical means. When this is done internal stress in the deposit may be revealed by a change of curvature of the coating from that originally imposed on it by contact with the substrate. Although it is sometimes possible to calculate the stress from the amount of curvature and the dimension of the sample (particularly in the case of cut tubular components) only approximate values can be achieved.

The second method of testing enables accurate measurements of internal stress to be made in the case of electrodeposited metal coatings. This is achieved by depositing the coating on to one side of a special, thin metal specimen and accurately measuring the deflection, the induced strain or the length of the composite so obtained. Flat sheet, flat or specially wound strip, or straight wire specimens secured in suitably instrumented holders have been proposed for tests described by Brenner and Senderoff[6], by Hoar and Arrowsmith[7], and by Dvorak and Vrobel[8]. Both compressive and tensile internal stresses may be accurately measured by these methods.

Brenner and Senderoff method. A helical strip specimen secured at one end is plated on one side only. Stress in the deposit causes contraction of the helix, and the movement of the free end is measured on a dial gauge; see *Figure 6.3(a)*.

Hoar and Arrowsmith method. A flat strip specimen secured at one end, and with a piece of soft iron wire secured to the top, is plated on one side only. Stress in the deposit causes the strip to bend, so deflecting a beam of light reflected from a mirror attached to the top of the strip. Current supplied to electromagnetic solenoids is used to measure the force required to overcome the bending due to deposit stress. See *Figure 6.3(b)*.

Dvorak and Vrobel method. A flat strip specimen, prestressed in a rigid framework, is plated on both sides. Stress in the deposit causes a change in length of the strip, measured with a dial gauge. See *Figure 6.3(c)*.

Ductility testing

As with internal stress, specially coated test specimens are normally required for testing ductility. The coating metal is deposited on to a soft ductile substrate such as polished brass and the specimen is bent either round a mandrel or on a spirally curved former (*Figure 6.4*). Ductility is assessed by the amount of elongation before cracking occurs calculated from the formula

$$E = \frac{100t}{D + t} \text{ per cent}$$

where t = total specimen thickness
D = diameter of curvature when cracking occurs

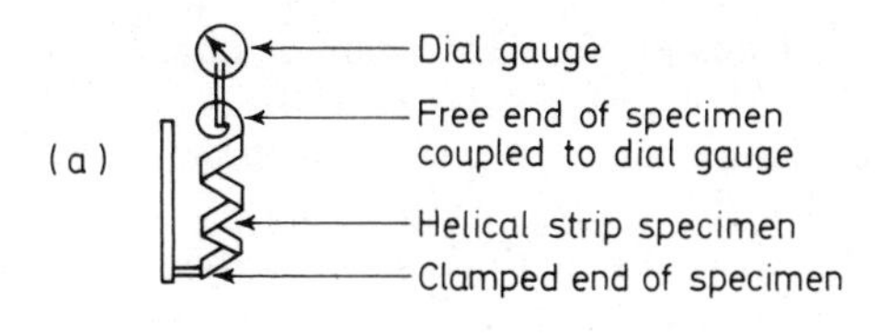

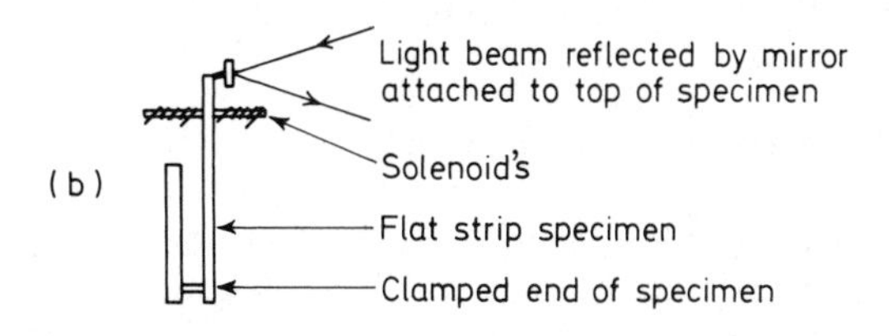

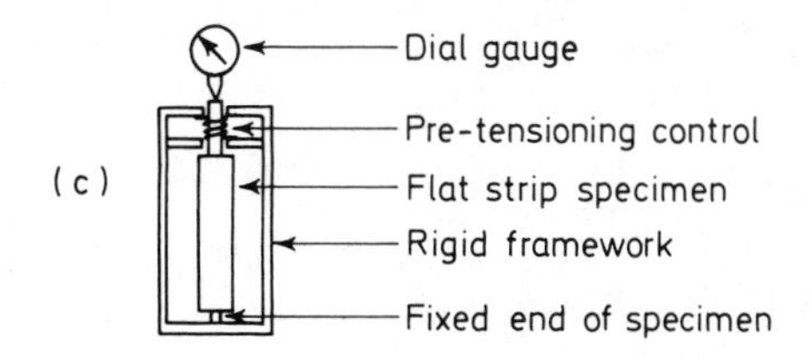

Figure 6.3 Methods of measuring stress in electrodeposits: (a) Brenner and Senderoff, (b) Hoar and Arrowsmith, (c) Dvorak and Vrobel

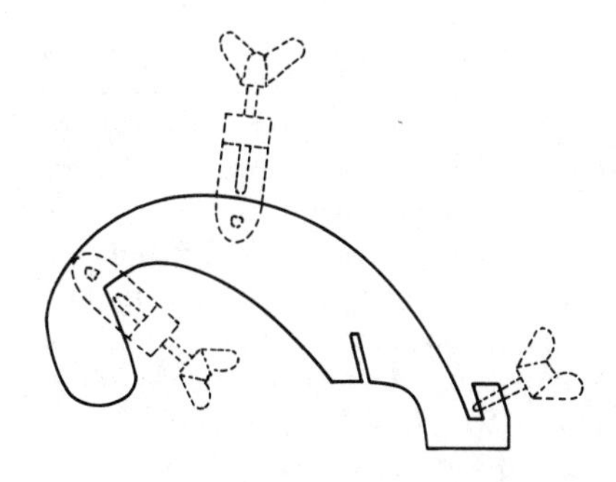

Figure 6.4 Edwards' bend-test former (after Edwards[9])

A modification of this method for use on detached coating foils is described in ASTM Method B490–68. A strip 6 × 75 mm in size is cut from the foil and its thickness measured with a micrometer. The strip is then bent into a 'U' shape and the legs of the 'U' are placed between the micrometer jaws; the jaws of the micrometer are then closed until cracks develop in the bent strip. The micrometer reading when cracking occurs is doubled in value and this figure is used as D in the ductility formula given above.

Tensile strength testing

The tensile strength of a metal coating, as also its ductility, can be determined by using a normal tensile testing machine to pull to destruction a test specimen, which may be either a coating detached from its substrate or one applied to a substrate having greater strength and ductility than that of the coating metal itself. Normal mechanical testing procedures apply in determining these coating properties when detached specimens are used, but when the coating is tested on a stronger substrate it is necessary to record the stress levels at which the coating fractures although the substrate of the specimen is still sound.

Hardness testing

The hardness of a metal coating may be determined by the normal hardness testing methods using either Vickers or Brinell indenting machines, but the results obtained are valid only when the thickness of the coating exceeds a critical minimum, which varies with the intrinsic hardness of the coating metal concerned. Below this thickness the value measured by the indenter is a composite, incorporating effects due to the hardness of both the coating and its substrate.

If it is required to measure the hardness of coatings thinner than the critical minimum it is necessary to use micro-hardness techniques, with the indenter operating on the polished surface of a mounted cross-section. In this way it is also possible to measure the different hardnesses of various components of an alloyed coating or of a multilayer coating system, although care must be taken to ensure that the micro-hardness impressions are made sufficiently far from the edge of each coating or component of a multi-coat system to avoid inaccurate edge effects. It must always be remembered that the micro-hardness value is not necessarily identical with the bulk hardness of a metal, although in general the difference between the two values is small.

Wear resistance testing

Coatings for engineering purposes may need to be tested to ensure that their wear resistance is adequate for the service requirement. Although hardness is sometimes related to wear resistance this is not always the case since, while it is generally true that a harder material will wear away a softer one, the reverse situation can apply in special cases. Consequently all wear resistance tests, which generally consist of measuring the amount of damage imparted by rubbing a sample under fixed known load against a reference surface or by treating the sample with standard abrasives, can give only comparative and empirical results. It is necessary, therefore, to relate performance in the chosen test to comparative results obtained on materials of known performance in the service environment, or to rely on the results of field tests. Such field tests may be accelerated somewhat by maintaining maximum aggressivity of the wear conditions continuously. As with all other forms of accelerated testing, however, the results must always be interpreted with caution and in the light of known service behaviour of the materials being tested.

Corrosion resistance testing

If metal coatings are applied for corrosion control it follows that the most important aspect of testing them is testing their resistance to corrosion. The purpose of corrosion testing is twofold: (a) tests are required to determine the performance of a given coating system in a particular corrosive environment, and (b) suitable corrosion tests may be used to reveal defects in a coating that could lead to inferior performance in service.

The ideal corrosion test is, of course, exposure to the 'natural' environment that will be met with in service under the conditions that will then apply. However, as the coating system more perfectly fulfils its function of corrosion control so the period of breakdown under these conditions becomes inordinately long, and more rapid means of testing must be sought. Accelerated corrosion tests may be devised to hasten breakdown by maintaining maximum severity continuously, by altering temperature or humidity or by using a specially aggressive artificial corrosive environment. Although breakdown may be achieved by these means within perhaps days, hours or (in extreme cases) even minutes, acceleration may produce attack that is very different in character from service performance because of the complex nature of the corrosion process. Thus the prediction of service life and mode of breakdown from the results of accelerated corrosion tests is fraught with danger and should never be attempted unless correlation has been firmly established as a result of extensive field testing.

To sum up, therefore, there is no universally applicable substitute for service performance data — even when only a single environment is considered. Although experience allows a limited degree of correlation to be obtained between service performance and the application of a particular test method for one type of protective system, it does not follow that if new systems behave similarly in that accelerated test they will do likewise in service.

By far the broadest and oldest category of tests is the use of fogs or sprays. On the assumption that the primary corrosion process in the atmosphere relies on the presence of moisture to maintain galvanic action and the presence of dissolved salts to increase the conductivity of the electrolyte, it was logical to expect an acceleration of corrosion by ensuring a plentiful supply of electrolyte and by increasing its conductivity by adding salts to supply the aggressive chloride ion.

The neutral salt spray test method was first introduced by Capp in 1914. It was designed to reproduce the atmosphere that might be encountered near the ocean. Certainly the use of this method speeds up the corrosion process above that which applies to the natural environment, but it soon became apparent that the results obtained did not correlate well with marine exposure performance and even less so with exposure to other types of atmospheres such as those polluted with sulphur contaminants.

This shortcoming is well illustrated by the way in which the neutral salt spray shows cadmium to be superior to zinc for the protection of steel. It is well known that zinc gives much better protection than cadmium when exposure to industrial atmospheres is involved, and even in marine environments the better of the two coatings varies with local environmental conditions. The reasons for these apparent anomalies are probably associated with the differing natures and solubilities of the corrosion products formed in the different types of environment. The provision of a plentiful supply of well-conducting electrolyte, such as is present in fog tests, suppresses any stifling action by corrosion products that may occur when drying out and rewetting occurs naturally, and in addition it overemphasises the 'effective throw' of the sacrifical protection obtained with anodic coatings of this type.

Over the years these limitations of the tests have become well recognised — to such an extent that warnings against their use for anodic coatings are printed in the ASTM specifications for zinc and cadmium coatings on steel.

When cathodic coatings are exposed to salt fog tests the stimulation of galvanic action that occurs accelerates attack at the points where breakdown first occurs; there is then a danger of the rest of the coated surface being protected, and fewer points of breakdown may thus be produced than is the case in natural exposure. Experience has shown that this is indeed the case, so a false picture is obtained of the corrosion behaviour in service. Appreciation of this point led to the search for modifications of the neutral salt spray test that would increase its severity (and so produce a pattern of corrosion more nearly resembling that occurring outdoors) and at the same time shorten the test period, which, at periods involving 100 or more hours, constituted another major limitation of its effective use. As these improved modified test methods became available and proven, the eventual abandonment of the neutral salt spray test for specification purposes for metal coatings became inevitable.

Although during the inter-war years the neutral fog tests were widely used, considerable research work was undertaken under the auspices of the ASTM and by many other workers in different countries. Variations in test parameters such as salt concentration, continuity of spray, temperature and relative humidity were all studied, but with little significant effect on the results obtained. It is interesting to speculate on the reasons for this lack of selectivity. Undoubtedly some part of the reason lies in the fact that the neutral fog tests generally are of a very mild nature, producing only a limited degree of corrosion in all but the least corrosion-resistant materials. Consequently, with a low overall degree of attack the effects of varying minor components of the system are difficult to detect. Apart from this, however, there is perhaps a more basic reason, related to the physical property of salts known as 'critical relative humidity'. When saturated solutions of salts are exposed in a closed volume a critical relative humidity is established and evaporation or condensation takes place to maintain that equilibrium. The operation of this mechanism in a fog test cabinet could easily nullify the effects of deliberate variations in spray rate or concentration over a very wide range. This is, perhaps, a similar mechanism to that which tends to occur in service when evaporation or drying out can and does occur, but obviously will bear very little direct relationship to the naturally occurring condition since in that case there is no closed system where equilibrium can be easily obtained, and furthermore wind, rain or movement can sweep away the corrodents to varying extents.

Apart from the continuous neutral salt spray test (currently defined in ASTM Method B117–64) there have been a number of variations proposed in which the salt is sprayed intermittently. The most widely used of these intermittent salt droplet tests is that specified in British Standard 1391 in which the solution for spraying consists of artificial sea water. Specimens are treated with an atomised spray of artificial sea water until a pattern of discrete droplets is produced on their surfaces, care being taken to ensure that the droplets are not allowed to coalesce and produce a complete film of moisture on the test surfaces. After spraying, the specimens are placed in a chamber in which the relative humidity approaches 100 per cent (obtained by having open vessels of water in the base of the chamber). They are removed for examination and respraying once each working day to ensure that the droplets do not completely dry out at any time throughout the test.

This type of test is of particular value for assessing the quality of passivation coatings on metals (see British Standard 3189) and for those metal coatings that are intended for use only in the milder environments. Its severity is not sufficient for testing the more corrosion-resistant coating systems.

A number of investigators studied acidification of the salt spray. Swindon and Stevenson suggested adding sulphuric acid to the sodium chloride used in an intermittent spray test, presumably attempting to introduce the sulphate ion, which is present in industrial environments, but this test method does not appear to have been extensively used. Nixon proposed acidification of continuous salt spray with acetic acid in 1945 as a result of a development programme carried out under ASTM auspices. The acetic acid salt spray test, which was finalised from this study, utilises a continuous spray of 5 per cent sodium chloride acidified with acetic acid to pH 3.2 and operated in a cabinet maintained at 35°C. Experience has shown that this test discriminates fairly well between different qualities of nickel + chromium coatings and reproduces satisfactorily the pattern of corrosion that occurs outdoors, although some anomalies in the results may occur when micro-discontinuous chromium systems are tested. The testing period is only slightly accelerated, and the requirement of a testing period of 8–114 hours constitutes a considerable limitation on the usefulness of the test. Nevertheless, the method is still in wide use, is included in most national standards and is favoured over other methods by many workers. (See British Standard 1224.)

The ASTM development programme that led to the production of the acetic acid salt spray was subsequently continued with a study of the contaminants present in rainwater in the Detroit area of the US and in the road wash picked up by cars operating in that region. A wide range of salts were detected and their addition, either singly or in combination, to accelerated test reagents was tried. From these many tests it was found that the addition of cupric chloride to the solution used in the acetic acid salt spray test produced a marked increase in the severity of the test while retaining the typical service corrosion patterns. Nixon and his colleagues finally defined a test procedure in 1956, consisting of the addition of 0.26 g/l cupric chloride to the acetic salt solution and increasing the testing temperature to 50°C. This test was subsequently adopted for standards purposes, being known as the CASS (copper-accelerated acetic acid salt spray) test. The original claims made for the method were of correlation between the results of a 16 hour test and one year's vehicle service in Detroit, and correlation has since been established between an 18 hour test and one year's industrial atmospheric exposure in the UK. A sketch of a cabinet suitable for the acetic salt spray or CASS test is shown in *Figure 6.5*.

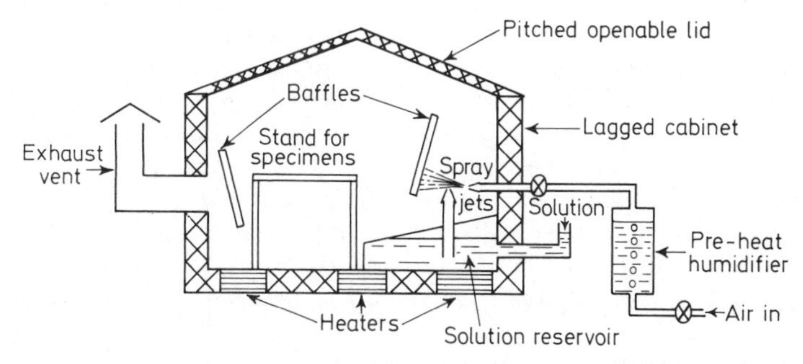

Figure 6.5 Cabinet for salt spray tests (acetic salt or CASS)

The CASS test is now universally accepted and is probably used more widely than any other fog test for control of the quality of nickel + chromium plated articles (see British Standard 1224). Testing times cover the range 8–24 hours according to the quality of the product, and many workers make use of multiple-cycle exposures for the more highly corrosion-resistant coating systems. However, a word of warning is necessary concerning the reliance that may be placed upon the results of these multiple CASS cycles — or indeed extended exposure to any accelerated test conditions. It does not necessarily follow that if, say, resistance to 18 hours in an accelerated test has been shown to correspond broadly with one year's service then resistance to 180 hours in that test would equate with 10 years' service. Extrapolation of correlation data is rarely valid. Even when using the CASS test within the limits of established correlation data care must be taken when interpreting the results obtained, particularly when new coating systems are being tested. Thus duplex nickel systems tend to be over-favoured in the CASS test compared with their benefits in many natural environments, and lateral spread of corrosion and the degree of sacrificial protection of the lower nickel layers may be exaggerated. Similarly, surface dulling develops on micro-discontinuous chromium deposits when they are CASS tested, to an extent and in a form that bears little or no resemblance to the surface dulling that sometimes occurs in severe natural environments. The rate of lateral spread of corrosion pits is markedly greater in CASS than in outdoor exposure when plated plastics are tested, and this must be compensated for in the correlation factors employed for these materials. All these anomalies in CASS testing stem from the provision of a plentiful supply of well-conducting electrolyte, which stimulates galvanic action, and the absence of alternate wetting and drying cycles, which would enable corrosion products *in situ* to stifle continued attack.

It is also necessary to operate the CASS test with very careful control of the variables if good reproducibility is to be obtained, both as regards different batches tested in an individual cabinet and as between tests carried out in different cabinets. Of all the variables concerned probably the most critical are the pH and the salt content of the collected spray; the volume of the spray is much less critical. A further very important factor that is often overlooked is the condition of the surface under test with respect to the method used for cleaning prior to testing, since both this method and the effectiveness of its application markedly influence the dropwise condensation of the fog on the specimens.

The contaminating salts found to be present in the road wash on vehicles were also tried as corrodents added to a paste that could be applied to the surface of a coated article instead of as a liquid to be sprayed on to its surface. A test procedure resulting from these studies was defined by Bigge in 1959 and, known as the Corrodkote test, has since become an accepted specification test for assessing the quality of plated metal articles. A mixture of cupric nitrate, ferric chloride and ammonium chloride is added to a kaolin-in-water paste, applied to the plated surface and, after being allowed to dry out, exposed for 20 hours to 95 per cent relative humidity. When used with nickel + chromium plated articles the pattern of corrosion produced has been found to correspond with that occurring on similarly plated articles in service for one year on motor vehicles in Detroit.

Best results with the Corrodkote test are obtained for plated steel articles. Any points of penetration to the basis metal are revealed as brown stains in the white paste coating. Corrosion of nickel or copper layers produces green or dark brown stains, which define cracks or pinholes in a chromium topcoat. With plated zinc alloy articles, however, the white corrosion products of zinc do not show up well and the corrosion blisters that are a feature of the service performance of this type of plated article are not produced in this test. One advantage of the Corrodkote test is that corrosion can be induced uniformly over the whole surface of intricately shaped articles, whereas all the fog type tests produce corrosion only on those portions of the surface that can be exposed to the fog over a limited range of angles of inclination so that the sprayed liquid can freely settle on those surfaces. The Corrodkote test is also detailed in British Standard 1224.

Although the utilisation of the Corrodkote test for specification purposes is limited to the detection of inferior quality in localised areas by the development of points of basis-metal corrosion, informed observation of the tested specimens can often yield a wealth of additional information. Thus, the location and distribution of green corrosion products can provide evidence of micro- or macro-cracks or pores in a chromium deposit applied over nickel. The presence or absence of a copper undercoat in a plating system may be revealed by copper corrosion products, and stress cracks penetrating partly or wholly through multi-layer coating systems may also be revealed.

Parallel with the development of accelerated salt fog tests studies were made of the use of the sulphate ion as an accelerating agent, having consideration of the fact that this is the effective ion present in industrially polluted environments. Thus, in the 1930s Evans and Britten advocated the use of weak sulphuric acid sprays, and Vernon suggested dilute sulphurous acid mixed with ammonium sulphate and used either with or without soldium chloride. Little further has been heard of the use of sulphuric acid as a corrodent spray, but the use of sulphurous acid has continued and has led to the well-known CRL sulphur dioxide test which uses the vapour from a sulphurous acid solution in a high-humidity environment. The Kesternich test, widely used at one time on the Continent of Europe for testing plated articles but now primarily used for paint coatings, also uses conditions similar to those of the CRL test, and the sulphur dioxide test developed by Edwards in 1958 was included for some years in British Standard 1224.

The sulphur dioxide test employs 0.5–2 per cent gaseous sulphur dioxide concentration in a relative humidity greater than 95 per cent for a period of 24 hours at 25°C. When used for plated articles it reveals discontinuities in chromium deposits by the green or brown corrosion products produced by attack on underlying layers of nickel or copper. However, the attack on these layers is excessive, tends to exaggerate the discontinuities and, in the case of micro-discontinuous chromium deposits, causes complete shattering of the chromium layer. Rusting of steel substrates is clearly revealed, but corrosion of zinc alloy substrates is not easily seen and blisters do not develop. Because of these limitations the test was dropped from British Standard 1224 in a later revision and now finds little or no favour in metal plating circles; its use is now largely confined to the detection of inadequate sealing of anodised aluminium such as would cause blooming in service, and in this field it performs extremely well.

The CRL and Kesternich tests may be used for detecting porosity in the thin precious-metal deposits used in the electrical and electronics industries, and mention has already been made of the rapid sulphur dioxide test developed by Clarke for this purpose (see page 153).

Various types of sulphur dioxide test are specified for testing different metal coatings. British Standard 1872 specifies 24 hours exposure at 20°C to air containing sulphur dioxide generated by the addition of one part of 0.1N sulphuric acid to four parts of a 10 g/l solution of sodium thiosulphate in a closed vessel for revealing porosity in tin coatings on steel; the same test is specified for quality control and porosity testing of 65/35 tin–nickel alloy coatings in British Standard 3597.

For gold coatings thicker than 5 μm British Standard 4292 specifies alternate exposure to an atmosphere containing 1 per cent sulphur dioxide injected into the chamber and 1 per cent hydrogen sulphide generated within the chamber by mixing sodium sulphide and 5 per cent sulphuric acid. For coatings thinner than 5 μm only exposure to the hydrogen sulphide atmosphere is required. Thin coatings of gold or silver may also be tested by exposure in a closed chamber containing thioacetamide vapour, produced by placing 0.3–0.5 g fine crystals of thioacetamide above a saturated solution of sodium acetate, which maintains the relative humidity within the chamber at 75 per cent.

A further, modern, adaption of sulphurous atmosphere tests is the industrial atmosphere test developed by Leeds and Such and specified in British Standard 2011 for environmental testing of coated components for electronic equipment. Exposure is for 20 hours at 25°C in an atmosphere containing 25 ppm sulphur dioxide and 3000 ppm carbon dioxide with a 75 per cent relative humidity. The conditions are established within the test chamber by injecting the products of controlled combustion of a hydrocarbon gas enriched with carbon disulphide, together with the appropriate quantity of air of controlled humidity, and ensuring between three and five complete changes of the test atmosphere within the cabinet every hour. The test is designed specifically for detecting minor changes in contact resistance due to superficial corrosion and has been shown to correlate well with service.

When testing those metal coatings that themselves actively corrode so as to provide sacrificial protection to the substrate (e.g. cadmium and zinc) it is often useful to employ very mild corrosion tests to provide information about the early stages of corrosion of the coating metal. For these purposes some form of exposure to high humidity without the acceleration provided by salt spray can be used. This may be straightforward exposure at a fixed temperature, either ambient or slightly elevated, perhaps with the specimens dampened with a distilled water spray before exposure, or cyclic exposure to differing conditions of temperature and/or relative humidity. Tests of this nature are often useful for detecting the proneness of zinc coatings to develop white rusting. A cyclic humidity test in British Standard 1706 requires exposure to 95 per cent relative humidity at 55°C for 16 hours, followed by five hours in the same humidity condition but at 30°C for each complete cycle.

A semi-accelerated test called the ASAP test was proposed by Goethner in 1970. This involves exposure to a natural polluted environment with the specimens contained in a louvred box open to the environment; the specimens are sprayed daily with natural sea water. It is said to be useful for developing the corrosion products that form on electrical contacts in service; the corrosion products may then be identified by analytical means. An exposure period of three weeks is employed and is claimed to represent extended normal service conditions.

Alternate immersion tests may be useful in that they may provide some of the effects of natural drying out in a reduced period, although there is considerable danger of effecting changes in the nature of the semi-solid corrosion products produced. An example of a test of this type is the 'dip and dry' test developed by General Motors for testing decoratively plated articles. The test article is immersed in a severely corrosive solution that simulates the road wash found on vehicles; this solution contains sodium sulphate, sodium sulphide and sodium thiosulphate together with chlorides of sodium and calcium, and the pH is adjusted to 9.3. After immersion in this solution for a period of only two seconds the test article is heated by infra-red lamps for a further 98 seconds; the dip and dry cycle is repeated over a period of four to eight hours.

Specially designed accelerated tests, often of a very complex nature, exist for application for particular specialised finishes. Perhaps the most complex of these is a cyclic test procedure described by Lascaro in the 1940s. The test was developed in the US for testing electronic components for military service applications, and an eight-stage test schedule is employed. After initial drying at 40°C there is a period of exposure at room temperature and 50 per cent relative humidity, followed by 95 per cent relative humidity at 65°C. These two conditions are repeated and an additional period of exposure to 50 per cent relative humidity at room temperature added, after which there is a period of cooling to −10°C followed by a final room temperature exposure to 95 per cent relative humidity. The complexity of this test is such that the full schedule requires a period in excess of a full working week, but Lascaro claimed that a successful result in the test ensures the high degree of reliability essential in military communications equipment.

In 1954 Pierce and Walter Pinner described an electrochemical test particularly designed to indicate the corrosion behaviour of thin plated metal coatings that offer only a very limited period of protection in service. Specimens are made anodic by 0.3 V against a copper cathode in a 3 per cent sodium chloride solution buffered with Rochelle salt; the test period is several hours.

Another example of an electrochemical test is the EC test described by Saur and Basco in 1966. This is probably the latest of the most rapid accelerated corrosion tests that have been specially developed for nickel + chromium plated components on either steel or zinc alloy substrates. The test specimens are potentiostatically controlled at 0.3 V anodic to a calomel reference electrode in a solution containing sodium nitrate, sodium chloride, nitric acid and water. The anodic conditions are applied cyclically on a basis of one minute on, two minutes off, and the applied current density is limited to a maximum of 3.3 mA/cm^2. This maximum current density is chosen as representative of the limiting current density that occurs on nickel + chromium plated components in service. The test is claimed to produce corrosion pits in a pattern corresponding to that which occurs in service outdoors, with the time to penetrate to the substrate correlating with performance in the CASS test and hence also with outdoor service in Detroit. For plated steel components, 2.4 minutes EC test is said to equal 16 hours CASS test (or one year Detroit service). Penetration of pits to the substrate in the test is revealed by the use of indicator solutions; for steel, phenanthroline hydrochloride indicator may be incorporated in the EC test solution, or alternatively a separate indicator solution containing thiocyanate acidified with acetic acid and containing hydrogen peroxide may be used; for zinc substrates a separate indicator solution containing quinoline acidified with acetic acid is used. It is also claimed that the dimensions of the pits produced in the EC test may be measured using a calibrated microscope, and that the size and number of pits correlated with similar data for pits occurring in outdoor service. The test is not easy to perform satisfactorily without both complex equipment and considerable expertise by the operator, and so is not readily applicable to routine testing, but undoubtedly it is a very useful tool for research and development studies of the performance of these types of plated finishes, having the great advantage of extreme rapidity in obtaining results.

Apart from the special accelerated tests — which may perhaps be classified as tests in 'un-natural' environments — slower corrosion tests may be carried out in more 'natural' environments such as exposure outdoors or immersion in waters or other liquids that may be encountered in service.

Even when testing in these 'natural' environments precautions must be taken to ensure that the results obtained may be related to actual service conditions. It has already been mentioned in Chapter 1 that in every type of natural environment there is a wide range of variables that can affect the corrosion process, and at least some of these must be allowed for when planning a natural-exposure test programme. For example, the influence of minor environmental changes on the corrosion process can be seen by considering surface dulling of micro-discontinuous chromium and the size of corrosion pits on plated plastics when different types of outdoor exposure are considered. These differences can lead to false conclusions being drawn about service performance. In the early days of micro-discontinuous chromium development, static exposure test results showed that failure by surface dulling occurred fairly rapidly in severely polluted environments. As a result of this the efficacy of the systems was questioned, but in service on vehicles it has been found that this mode of failure rarely occurs within practical periods of service. Hence for these systems the severe static exposure test may be considered to be a form of accelerated test — albeit a very protracted one — and one that does not truly reproduce the type of breakdown that is of significance in service. Limited service trials so far carried out with plated plastics suggest that similar considerations also apply to the rate of deterioration by lateral spread of superficial corrosion pits, which is markedly greater in static exposure to polluted environments than in service on vehicles.

Table 6.2 lists some examples of the types of variable that need careful consideration when designing tests in the atmosphere or immersed in liquids. This list is by no means exhaustive and other

Table 6.2 FACTORS AFFECTING THE SEVERITY OF CORROSION TESTS

Atmospheric tests	*Immersion tests*
Angle of exposure of specimens	Composition of electrolyte
Meteorological factors	Temperature
Atmospheric pollution	Degree of aeration
Static or mobile	Stagnant or flowing
Retention of dirt	Presence of shielding deposits
Frequency of washing	Presence of entrained abrasives

variables relevant to specific service requirements will readily come to the minds of those concerned.

A final point, of the greatest importance in all corrosion testing, concerns the assessment of damage in the tests and the interpretation placed on the results. With coated metals one of the most widely used criteria for assessing corrosion, namely loss in weight, is unlikely to be of great use since the serviceability of a coating system is dependent upon limiting damage to the substrate and any such damage may well be unacceptable long before any significant weight loss affects the complete coated article.

Sometimes the extent of damage to the substrate can be measured by its effect on the mechanical properties of the complete article, and in such cases it is obviously simple to establish accurately quantified acceptance levels. Similar considerations apply where the effect of damage on physical properties such as conductivity and resistivity is concerned.

With most coating systems, however, damage needs to be assessed by the extent to which substrate corrosion products can be tolerated either on grounds of their effect on appearance or because they interfere with the use of the article by contamination. When these considerations apply it is generally necessary to set limits to the size and number of defects, or to assess the percentage area of defective coating surface. Assessments of this type are usually subjective, although some degree of qualitative assessment may be incorporated in suitable cases. Assessment is frequently made by visual comparison with standard charts of defects or of defective areas; since the adverse effects of corrosion on appearance are most important in the early stages of breakdown, these are usually compiled on a logarithmic basis of the extent to which breakdown has occurred.

Apart from the subjective nature of the actual assessment it is also necessary to set purely subjective acceptance levels in most practical cases, so considerable expertise is required in order to interpret results in the light of actual service requirements.

Appendix 1

British Standards relevant to coated materials

General series

BS 182	Galvanised line-wire for telegraph and telephone purposes
BS 183	General purpose galvanised steel wire strand
BS 215	Part 2: Aluminium conductors, steel reinforced
BS 365	Galvanised steel wire ropes for ships
BS 417	Galvanised mild steel cisterns and covers, tanks and cylinders
BS 443	Galvanised coatings on wire
BS 729	Hot-dip galvanised coatings on iron and steel articles
BS 801	Lead and lead alloy sheaths of electric cable
BS 1224	Electroplated coatings of nickel and chromium
BS 1391	Performance tests for protective schemes used in the protection of light-gauge steel and wrought iron against corrosion
BS 1441	Galvanised steel wire for armouring submarine cables
BS 1442	Galvanised mild steel wire for armouring cables
BS 1485	Galvanised wire netting
BS 1565	Galvanised mild steel indirect cylinders, annular or saddle-back type
BS 1689	Galvanised mild steel fire buckets
BS 1706	Electroplated coatings of cadmium and zinc on iron and steel
BS 1822	Nickel clad steel plate
BS 1872	Electroplated coatings of tin
BS 2011	Methods for the environmental testing of electronic components and electronic equipment
BS 2569	Sprayed metal coatings
BS 2816	Electroplated coatings of silver for engineering purposes

BS 2920	Cold-reduced tinplate and cold-reduced blackplate
BS 2989	Hot-dip galvanised plain steel sheet and coil
BS 3034	Galvanised hollow-ware
BS 3083	Hot-dipped galvanised corrugated steel sheets for general purposes
BS 3189	Phosphate treatment of iron and steel
BS 3315	Watch case finishes in gold alloys
BS 3382	Electroplated coatings on threaded components
BS 3393	Tinned steel baking dishes
BS 3597	Electroplated coatings of 65/35 tin-nickel alloy
BS 3654	Galvanised steel dustbins for dustless emptying
BS 3740	Steel plate clad with corrosion-resisting steel
BS 3745	Evaluation of results of accelerated corrosion tests on metallic coatings
BS 3788	Tin coated finish for culinary utensils
BS 4025	The general requirements and methods of test for printed circuits
BS 4087	Copper-covered steel wire for telephone and telegraph purposes
BS 4290	Electroplated coatings of silver for decorative purposes on nickel, silver and copper
BS 4292	Electroplated coatings of gold and gold alloy
BS 4393	Tin or tin-lead coated copper wire
BS 4479	Recommendations for the design of metal articles that are to be coated
BS 4495	Recommendations for the flame spraying of ceramic and cement coatings
BS 4584	Metal clad base materials for printed circuits
BS 4601	Electroplated coatings of nickel plus chromium on plastics materials
BS 4641	Electroplated coatings of chromium for engineering purposes
BS 4758	Electroplated coatings of nickel for engineering purposes
BS 4761	Sprayed unfused metal coatings for engineering purposes
BS 4921	Sheradised coatings on iron and steel articles
BS 4950	Sprayed and fused metal coatings for engineering purposes

Codes of Practice

CP 143	Sheet roof and wall coverings, Parts 2 & 10: Galvanised corrugated steel
CP 2008	Protection of iron and steel structures from corrosion
CP 3012	Cleaning and preparation of metal surfaces

Aerospace series

2A59	Cadmium-plated steel hexagonal-headed bolts with close tolerance shanks for aircraft
2A60	Cadmium-plated steel hexagonal-headed shear bolts for aircraft
3A111	Cadmium-plated steel bolts with close tolerance shanks for aircraft
2A112	Cadmium-plated shear bolts
3L72	Aluminium-coated sheet and strip of aluminium-copper-magnesium-silicon-manganese alloy (solution treated and aged at room temperature)
3L73	Aluminium-coated sheet and strip of aluminium-copper-magnesium-silicon-manganese alloy (solution treated and precipitation treated)
2L88	Aluminium-alloy-coated sheet and strip of aluminium-zinc-magnesium-copper-chromium alloy (solution treated and precipitation treated)
2L89	Close toleranced sheet and strip of aluminium-coated aluminium-copper-magnesium-silicon-manganese alloy (solution treated and aged at room temperature)
2L90	Close toleranced sheet and strip of aluminium-coated aluminium-copper-magnesium-silicon-manganese alloy (solution treated and precipitation treated)
L107	Aluminium-coated sheet and strip of aluminium-copper-magnesium-silicon-manganese alloy (supplied for solution treatment by the user)
L108	Close toleranced sheet and strip of aluminium-coated aluminium-copper-magnesium-silicon-manganese alloy (supplied for solution treatment by the user)
L109	Aluminium-coated sheet and strip of aluminium-copper-magnesium-manganese alloy (solution treated and aged at room temperature)
L110	Aluminium-coated sheet and strip of aluminium-copper-magnesium-manganese alloy (supplied for solution treatment by the user)
SP113	Cadmium plated close tolerance shear pins for aircraft
2W9	Preformed galvanised carbon steel wire rope

STA specification

STA 23	Terneplate (tin-terne) quality

Draft for Development

DD 24	Methods of protection against corrosion on light section steel used in building

Appendix 2

ASTM Standards and Test Methods relevant to coated materials

A90-69	Method of test of galvanised coating weight
A112-66	Zinc-coated (galvanised) steel tie wire
A116-66	Zinc-coated (galvanised) fencing
A121-69	Zinc-coated (galvanised) barbed wire
A123-69	Zinc (hot-galvanised) coatings on products fabricated from rolled, pressed and forged steel shapes, plates, bars and strip
A153-67	Zinc-coated (galvanised) hardware
A163-36	Zinc-coated (galvanised) wrought iron sheet
A164-55	Electrodeposited coatings of zinc on steel
A165-55	Electrodeposited coatings of cadmium on steel
A263-66, A264-66, A265-70	Clad steel plates for corrosion resistance
A308-69	Terne-coated cold rolled steel sheet.
A309-54	Triple-spot test method for weight and composition of coating on terne sheet
A361-67	Galvanised iron or steel roofing sheet
A386-67	Hot dip zinc coated assembled products
A394-65	Bolts and nuts galvanised for transmission towers
A444-67	Galvanised sheet for culverts and under drains
A446-69	Galvanised sheet of structural quality. Coil and cut lengths
A463-69	Cold rolled aluminium-coated steel sheet
A526-67	Galvanised carbon steel. Commercial quality
A527-67	Galvanised carbon steel sheet. Lock-forming quality
A528-67	Galvanised carbon steel rimmed sheet. Drawing quality
B117-64	Salt spray (fog) testing

B177-68	Engineering coatings of chromium
B188-49	Practice of preparation of low carbon steel for electroplating.
B200-60	Electrodeposited coatings of lead on steel
B201-68	Testing chromate coatings on zinc and cadmium
B242-54	Preparation of high carbon steel for electroplating
B252-54	Preparation of zinc alloy diecastings for electroplating
B253-68	Preparation and plating on aluminium alloys by zincate
B254-53	Preparation and plating on stainless steel
B281-58	Preparation of copper and its alloys for electroplating
B287-62	Acetic acid salt spray test
B319-60	Preparation of lead and its alloys for electroplating
B320-60	Preparation of iron castings for electroplating
B322-68	Cleaning metals prior to electroplating
B343-67	Preparation of nickel for electrodeposition of nickel
B368-68	CASS test
B380-65	Corrodkote test
B454-70	Mechanically deposited cadmium and zinc on ferrous materials
B456-67	Electrodeposition of nickel plus chromium
B480-68	Preparation of magnesium and its alloys for electroplating
B481-68	Preparation of titanium and its alloys for electroplating
B482-68	Preparation of tungsten and its alloys for electroplating
B487-68	Microscope thickness test
B488-71	Electrodeposited coatings of gold for engineering purposes
B489-68	Ductility bend test
B490-68	Micrometer bend test
B499-60	Magnetic thickness testing
B504-70	Coulometric thickness testing
B507-70	Design of articles for rack plating
B529-70	Eddy current thickness testing
B530-70	Magnetic thickness testing
B533-70	Evaluation of appearance of plated plastics
B537-70	Rating of electroplated panels subject to atmospheric exposure tests
B545-71	Electrodeposited coatings of tin
B553-71	Thermal cycling test
B554-71	Measurement of thickness of metallic coatings on non-metallic substrates
F1-68	Nickel clad and nickel plated steel strip for electron tubes

F2-68		Aluminium clad steel strip and nickel steel aluminium composite strip for electron tubes
STP403	1966	Cold cleaning with halogenated solvents

References

1. Uhlig, H.H., *Corrosion,* **6,** 29–33 (1950)
2. Carter, V.E., 'The effect of painting over sprayed metal coatings on aluminium alloys', *J. Inst. Metals,* **91,** 413 (1963)
3. *The BNF jet test for local thickness measurement of electrodeposited metallic coatings,* British Drug Houses Ltd, Poole, Dorset
4. White, R.A., 'Coulometric plating thickness meter', *Metal Industry,* **98**(23), 455 (1961)
5. Clarke, M. and Sansum, A.J., 'A two-hour porosity test for gold on substrates of copper, silver and nickel', *Trans. Inst. Metal Finishing,* **50**(5), 211 (1972)
6. Brenner, A. and Senderoff, S., 'A spiral contractometer for measuring stress in electrodeposits', *J. Res. Natn Bur. Standards,* **42,** 89 (1949)
7. Hoar, T.P. and Arrowsmith, D.J., 'Stress in nickel electrodeposits', *Trans. Inst. Metal Finishing,* **36,** 1 (1958)
8. Dvorak, A. and Vrobel, L., 'A new method for the measurement of internal stress in electrodeposits', *Trans. Inst. Metal Finishing,* **49,** 153 (1971)
9. Edwards, J., 'Spiral bending test for electrodeposited coatings'. *Trans. Inst Metal Finishing,* **35,** 101 (1958)

Bibliography

Shreir, L.L. (ed), *Corrosion,* 2nd edn (2 vols), Newnes–Butterworths (1976)

Burns, R.M. and Bradley, W.W., *Protective coatings for metals,* 3rd edn, Reinhold (1967)

Dennis, J.K. and Such, T.E., *Nickel and chromium plating,* Newnes–Butterworths (1972)

Pollack, A., Westphal, P. and Weiner, R., *An introduction to metal degreasing and cleaning,* R. Draper Ltd (1963)

Plaster, H.J., *Blast cleaning and allied processes,* Industrial Newspapers Ltd (1972)

Burhart, W., Silman, H. and Draper, C.R., *Polishing,* R. Draper Ltd (1960)

Silman, H., *Chemical and electro-plated finishes,* Chapman and Hall (1952)

Ballard, W.E., *Metal spraying,* 4th edn, Griffin (1963)

Smart, R.F. and Catherall, J.A., *Plasma spraying,* Technical Library TL/ME/3, Mills and Boon (1972)

Powell, C.F., Campbell, I.E. and Gonser, B.W., *Vapour plating,* J. Wiley and Chapman and Hall (1955)

Muller, G. and Baudrand, D.W., *Plating on plastics,* 2nd edn, revised by G.D.R. Jarrett and C.R. Draper, R. Draper Ltd (1971)

Champion, F.A., *Corrosion testing procedures,* 2nd edn, Chapman and Hall (1964)

Reid, F.H. and Goldie, W., *Gold plating technology,* Electrochemical Publications Ltd (1974)

Handbook on electroplating, W. Canning & Co. Ltd, 21st edn (1970)

Lowenheim, F. (ed.), *Modern electroplating,* 2nd edn, J. Wiley (1963)

Index